Contemporary's NUMBER POWER

Introductory

CB

CONTEMPORARY BOOKS

a division of NTC/Contemporary Publishing Group
Lincolnwood, Illinois USA

ISBN: 0-8092-0609-9

Published by Contemporary Books,
a division of NTC/Contemporary Publishing Group, Inc.,
4255 West Touhy Avenue,
Lincolnwood (Chicago), Illinois 60646-1975 U.S.A.

Manufactured in the United States of America.

890 CU 0987654321

Contents

To the Learner

Even if math has never been easy for you, this text will give you the instruction and practice you need to understand the basics. In this global and technological society, an understanding of math is important, and, at some time or another, you will be asked to demonstrate that you can solve math problems well.

Using *Number Power Introductory, Level E* is a good way to develop and improve your mathematical skills. It is a comprehensive text for mathematical instruction and practice. Beginning with basic comprehension skills in addition, subtraction, multiplication, and division, *Number Power Introductory, Level E* includes basic concepts about performing computations. For instance, concepts include information such as words like *plus, sum,* and *total* often signal addition problems; the addition symbol is a "+" sign, and addition problems have distinctive characteristics. Then problems, which illustrate the basic kinds of addition problems learners will confront, are included for plenty of practice.

Accompanying all of this practice are a Skills Inventory Pre-Test and a Skills Inventory Post-Test. The Skills Inventory Pre-Test will help you identify your math strengths and weaknesses before you begin working in the book. Then you can work in those areas where additional instruction and practice are needed. Upon completion of these exercises, you should take the Skills Inventory Post-Test to see if you have achieved mastery. Mastery is whatever score you and your instructor have agreed upon to be correct to insure that you understand each group of problems.

Usually mastery is completing about 80 percent of the problems correctly. After achieving mastery, you should then move on to the next section of instruction and practice. In this way the text offers you the chance to learn at your own pace, covering only the material that you need to learn. In addition, the instruction on each page offers you the opportunity to work on your own.

Addition, subtraction, multiplication, and division instruction and practice are presented in the first part of *Number Power Introductory, Level E*. The other part contains applications. It includes numeration, number theory, data interpretation, pre-algebra, measurement, geometry, and estimation. You will work with ordinal numbers, place value of numbers, graphs, tables, charts, number sentences, calendars, time, plane figures, logical reasoning, problem solving, estimation, and many other topics.

Completing *Number Power Introductory, Level E* will make you more confident about doing mathematical problems. Remember to use the Answer Key in the back of the book to check your responses. Soon you will find yourself either enjoying math for the first time or liking it even more than you did previously.

Skills Inventory Pre-Test

Part A: Computation

Circle the letter for the correct answer to each problem.

1

$$\begin{array}{r} 46 \\ +32 \\ \hline \end{array}$$

A 13
B 74
C 78
D 80
E None of these

2

$$\begin{array}{r} 98 \\ -86 \\ \hline \end{array}$$

F 11
G 12
H 13
J 14
K None of these

3 $7 \times 7 =$ ____

A 49
B 56
C 77
D 14
E None of these

4 $6 \div 6 =$ ____

F 6
G 1
H 12
J 2
K None of these

5

$$\begin{array}{r} 84 \\ +49 \\ \hline \end{array}$$

A 123
B 35
C 122
D 133
E None of these

6

$$\begin{array}{r} 203 \\ \times \quad 3 \\ \hline \end{array}$$

F 233
G 209
H 600
J 290
K None of these

7 $130 - 10 =$ ____

A 110
B 120
C 30
D 129
E None of these

8 $48 \div 2 =$ ____

F 22
G 42
H 26
J 44
K None of these

9 $7 \div 2 =$ ____

A 2
B 3 r 1
C 3
D 2 r 1
E None of these

10

$$\begin{array}{r} 32 \\ 42 \\ +\ 6 \\ \hline \end{array}$$

F 76
G 70
H 80
J 82
K None of these

11

$$\begin{array}{r} 432 \\ -\quad 6 \\ \hline \end{array}$$

A 436
B 434
C 424
D 426
E None of these

12

$$\begin{array}{r} 15 \\ \times\ 8 \\ \hline \end{array}$$

F 115
G 50
H 120
J 125
K None of these

Part B: Applied Mathematics

Circle the letter for the correct answer to each problem.

13 Which of these has the same answer as the problem in the box?

1 + 2 + 3

A 3 − 3
B 3 + 3
C 1 × 5
D 2 + 3

14 Which of these is a set of odd numbers?

F 1, 3, 5, 9
G 1, 2, 3, 4
H 5, 10, 15, 20
J 2, 4, 6, 8

15 Which sign makes this number sentence true?

15 □ 5 = 10

A +
B ×
C −
D ÷

16 A 4-hour train ride begins at 10:00 A.M. When will it end?

F 2:00 P.M.
G 4:00 P.M.
H 5:00 P.M.
J 9:00 A.M.

Read this passage. Then do Numbers 17 through 19.

Cesar is driving to Toledo. He will use a rented car.

17 When Cesar begins driving, the mileage odometer reads 1,802. Which of these is 1,802?

A one thousand, eight hundred two
B one thousand, eight hundred twenty
C one hundred eighty-two
D ten thousand, eight hundred two

18 There are four routes Cesar could take. Which one is the shortest?

F 864 miles
G 803 miles
H 796 miles
J 850 miles

19 When Cesar has driven half way, his gas gauge looks like this:

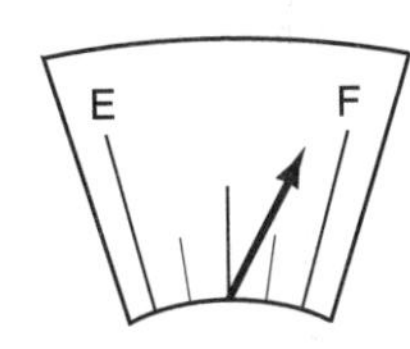

What fraction of his gas tank is filled?

A $\frac{1}{2}$ **C** $\frac{1}{3}$

B $\frac{3}{4}$ **D** $\frac{1}{4}$

20 Which numbers are missing from the number sequence?

24, 28, ?, 36, 40, ?

F 29 and 41
G 30 and 42
H 31 and 43
J 32 and 44

21 A machine takes each "In" number and produces the "Out" number. Which of these could be the rule that changes each "In" number to an "Out" number?

In	5	3	6	9
Out	2	0	3	6

A Subtract 1.
B Subtract 2.
C Subtract 3.
D Subtract 4.

22 Art bought a book for \$8.99 plus 72¢ tax. How much did he pay in all?

F \$8.61
G \$8.71
H \$9.61
J \$9.71

23 Ray spent \$15.15 for a shirt and \$40.15 for a pair of pants. He paid \$2.35 tax for his purchase. How much did he spend in all?

A \$57.65
B \$57.75
C \$58.65
D \$58.75

24 A doorway is 24 inches wide. Which measurement is the same as twenty-four inches?

F 2 feet
G $2\frac{1}{4}$ feet
H 6 feet
J 12 feet

Mrs. Owens used this graph to record how much housework her family did one week. **Study the graph. Then do Numbers 25 through 27.**

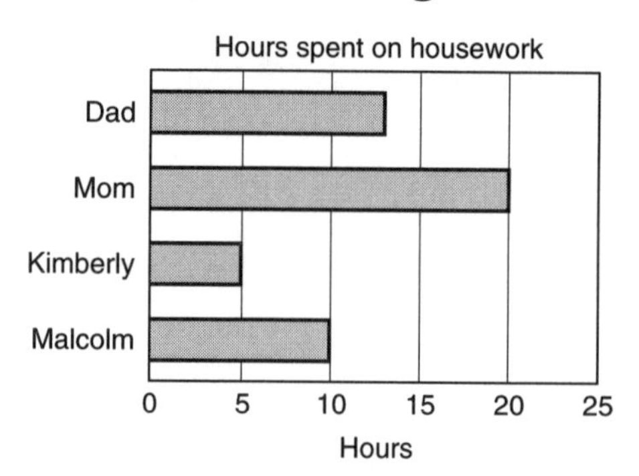

25 Who spent the most time doing housework?

A Dad
B Mom
C Kimberly
D Malcolm

26 How much time did Malcolm spend doing housework?

F 5 hours
G 10 hours
H 15 hours
J 20 hours

27 What is the average amount of time each family member spent on housework? Pick the best estimate.

A 6 hours
B 9 hours
C 7 hours
D 12 hours

28 What temperature is shown on this thermometer?

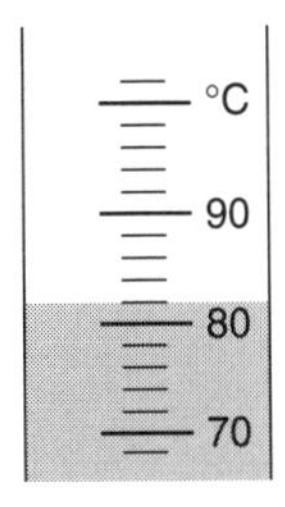

F 80°C
G 81°C
H 82°C
J 84°C

29 Glenn pays a $15.00 school activity fee for each of her 3 children. How much does she pay altogether?

A $45.00
B $65.00
C $75.00
D $90.00

30 Suli earns $10.00 an hour when she works overtime. How much does he earn for 8 hours of overtime?

F $ 8.00
G $ 10.00
H $ 80.00
J $800.00

Use this ad to answer questions 31 through 34.

Chef's Pantry Spring Clearance
Saturday and Sunday only!

Ceramic Mixing Bowl	$22.50
Hand Mixer	$12.95
Marble Rolling Pin	$18.19
Pastry Bag	$5.45
Serving Platter	$42.85
Cookie Cutter	$1.10
Wagner Steak Knives (4-piece set)	$15.95
8-inch cake pan	$6.50
4-inch cake pan	$4.10

31 About how much would it cost to buy a mixing bowl and a pastry bag?

A $23.00 **C** $27.00
B $25.00 **D** $28.00

32 About how much does each steak knife cost?

F $4.00 **H** $5.00
G $4.50 **J** $5.50

33 About how much would it cost to buy a hand mixer, a serving platter, and a rolling pin?

A $65.00 **C** $75.00
B $70.00 **D** $80.00

34 Trey buys 3 cookie cutters with a $5.00 bill. About how much change should he get back?

F $1.50 **H** $3.00
G $2.50 **J** $3.50

35 Which of these shapes will make a square when it is folded along the dotted line?

A 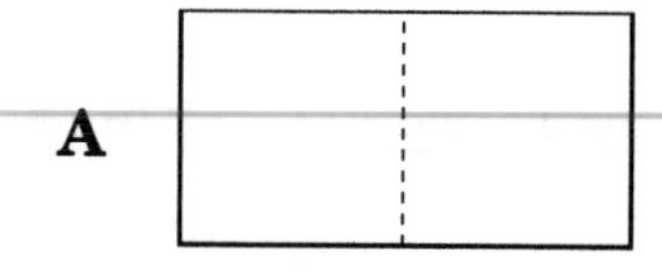

B 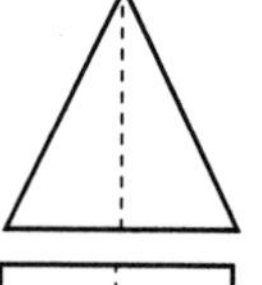

C

D 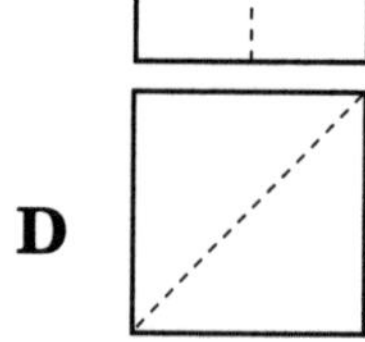

36 Which of these is the same size and shape as the dark figure?

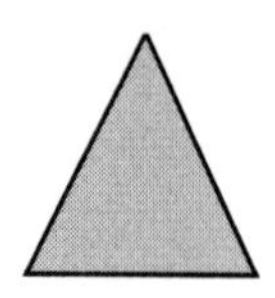

F 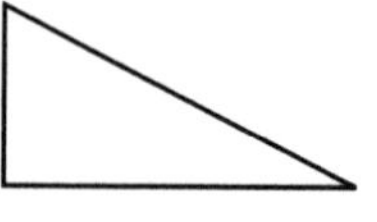

H 

G 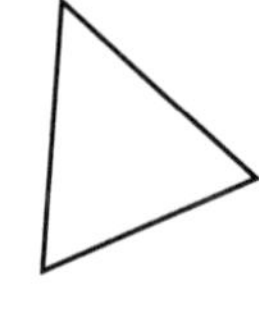

J 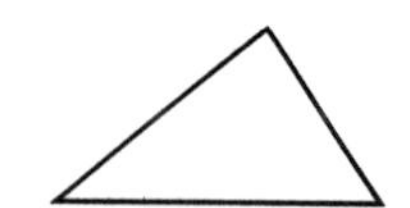

37 This box shows a repeating pattern of shapes. What will the next shape be?

A a square
B a rectangle
C a triangle
D a circle

Skills Inventory Pre-Test Evaluation Chart

Use the key to check your answers on the Pre-Test. The Evaluation Chart shows where you can turn in the book to find help with the problems you missed.

Key

1	C
2	G
3	A
4	G
5	D
6	K
7	B
8	K
9	B
10	H
11	D
12	H
13	B
14	F
15	C
16	F
17	A
18	H
19	B
20	J
21	C
22	J
23	A
24	F
25	B
26	G
27	D
28	H
29	A
30	H
31	D
32	F
33	C
34	F
35	A
36	G
37	D

Evaluation Chart

Problem Numbers	Skill Areas	Practice Pages
1, 5, 10	Addition	17–30
2, 7, 11	Subtraction	31–45
3, 6, 12	Multiplication	46–57
4, 8, 9	Division	58–72
17, 18, 19	Numeration	1–16
13, 14, 20	Number Theory	1–16, 31, 46, 58
25–27	Data Interpretation	73–85
15–21	Pre-Algebra	86–99
16, 24, 28	Measurement	100–118
35–37	Geometry	119–126
22, 23, 29, 30	Computation in Context	8, 26–28, 41–43, 53–55, 68–70
31–34	Estimation	11–14, 25, 39–40, 52, 67

Number Power Introductory, Level E

Correlations Between Number Power Introductory *and TABE™ Mathematics Computation*

Addition of Whole Numbers **Pre-Test Score** ☐ **Post-Test Score** ☐

Subskill	**TABE, Form 7 Item Numbers**	**TABE, Form 8 Item Numbers**	**Practice and Instruction Pages in This Text (p means practice page.)**	**Additional Practice and Instruction Resources**
Adding 1–3 digits without regrouping	1, 3	2, 7, 9	17, 18, 19p, 22, 29p, 30p	Number Power, Bk. 1, pages 6–9 *Breakthroughs in Math / Bk. 1,* pages 20, 21, 24–27 *Number Sense,* Bk. 1, pages 12, 14, 16, 17
Adding a column of 1–3 digits without regrouping	2		19–22, 29p, 30p	*Breakthroughs in Math / Bk. 1,* pages 24, 25 *Foundations: Mathematics,* pages 25, 26
Adding three 1-digit numbers		1	20, 29p, 30p	*Breakthroughs in Math / Bk. 1,* pages 22, 23 *Foundations: Mathematics,* pages 25, 26
Adding 1–3 digit numbers with regrouping	10, 11, 13	3, 8	23, 24p, 29p, 30p	*Number Power,* Bk. 1, pages 10, 11, 14 *Breakthroughs in Math / Bk. 1,* pages 32–34, 36, 37 *Number Sense,* Bk. 1, pages 19, 20
Adding a column of digits with regrouping	12	10	23, 24, 29p, 30p	*Number Sense,* Bk. 1, pages 21–23 *Foundations: Mathematics,* pages 25, 26 *The Math Problem Solver,* pages 10, 11

Corresponds to TABE™ Forms 7 and 8

Subtraction of Whole Numbers **Pre-Test Score** ☐ **Post-Test Score** ☐

Subskill	TABE, Form 7 Pages	TABE, Form 8 Pages	Practice and Instruction Pages in This Text	Additional Practice and Instruction Resources
Basic subtraction facts		4	31–33, 44p, 45p	*Number Power,* Bk. 1, pages 22–23 *Breakthroughs in Math / Bk. 1,* pages 46, 47 *Number Sense,* Bk. 1, pages 28, 29
Subtracting from 2 or 3 digits, without regrouping	4, 5, 6	5, 6, 14	34, 35, 44p, 45p	*Number Power,* Bk. 1, pages 24, 25 *Breakthroughs in Math / Bk. 1,* pages 48–50 *Foundations: Mathematics,* pages 30–32
Subtracting from 2 or 3 digits, with regrouping	16, 17, 18, 19	13, 15, 16	36–38, 44p, 45p	*Breakthroughs in Math / Bk. 1,* pages 56–60, 62–64 *Number Sense,* Bk. 1, pages 33–42, 44, 45 *Foundations: Mathematics,* pages 33–35

Multiplication of Whole Numbers **Pre-Test Score** ☐ **Post-Test Score** ☐

Subskill	TABE, Form 7 Pages	TABE, Form 8 Pages	Practice and Instruction Pages in This Text	Additional Practice and Instruction Resources
Basic Multiplication facts	8, 9	11, 17	46, 47, 48p, 56p, 57p	*Breakthroughs in Math / Bk. 1,* pages 74, 75 *Number Sense,* Bk. 2, pages 5–7 *Foundations: Mathematics,* pages 40, 41 *Critical Thinking with Math,* pages 3–5
Multiplying by 1 digit without regrouping	7, 20	12	49, 56p, 57p	*Number Power, Bk. 1,* pages 49, 52 *Breakthroughs in Math / Bk. 1,* pages 76, 81
Multiplying by 1 digit with regrouping	21, 22	18, 20	50, 51, 56p, 57p	*Number Power, Bk. 1,* pages 54, 55 *Breakthroughs in Math / Bk. 1,* pages 84–87, 91 *Foundations: Mathematics,* pages 49–52
Multiplying three 1-digit numbers		25	46, 47p, 56p, 57p	

Division of Whole Numbers   Pre-Test Score ☐   Post-Test Score ☐

Subskill	TABE, Form 7 Pages	TABE, Form 8 Pages	Practice and Instruction Pages in This Text	Additional Practice and Instruction Resources
Basic division facts	14, 15	21, 22, 24	58–60, 66p, 71p, 72p	*Breakthroughs in Math / Bk. 1,* pages 102, 103 *Number Sense,* Bk. 2, pages 24, 25 *Foundations: Mathematics,* pages 42,43
Dividing by 1 digit with no remainder	23	23	61–63, 65, 66p, 71p, 72p	*Number Power, Bk. 1,* pages 72–75 *Number Sense,* Bk. 2, pages 26, 28–33 *Foundations: Mathematics,* pages 59–66
Dividing by 1 digit with a remainder	24, 25	19	64, 65, 66p, 71p, 72p	*Number Power, Bk. 1,* pages 76–79 *Breakthroughs in Math / Bk. 1,* pages 108, 109 *Number Sense,* Bk. 2, pages 27–33

Correlations Between Number Power Introductory *and TABE Applied Mathematics*

Numeration **Pre-Test Score** ☐ **Post-Test Score** ☐

Subskill	TABE, Form 7 Pages	TABE, Form 8 Pages	Practice and Instruction Pages in This Text	Additional Practice and Instruction Resources
Ordinal Numbers		36	6, 15p, 16p, 73p, 77p, 84p, 85p, 87p	
Word names	27		3, 15p, 16p	*Number Power Review,* pages 2, 3 *Breakthroughs in Math / Bk. 1,* pages 9, 10
Recognizing numbers		41	4, 15p, 16p	*Number Power Review,* pages 2, 3 *Number Sense,* Bk. 1, page 6
Ordering numbers	2		5, 15p, 16p, 80p, 74–78p, 84p, 85p	*Number Sense,* Bk. 1, pages 8, 9
Place value	4	28, 40	1, 2p, 3, 15p, 16p	*Number Power,* Bk. 1, pages 1–4, 7 *Breakthroughs in Math / Bk. 1,* pages 8, 9 *Number Sense,* Bk. 1, pages 1–4, 7
Fractional part	31, 43		7–10, 15p, 16p, 76–85p	*Number Power,* Bk. 2, pages 5–6 *Number Sense,* Bk. 5, pages 1–13, 28–30, 55–59 *Critical Thinking with Math,* pages 78, 79
Properties	5		17, 18p, 31, 32p, 46, 47p, 58, 59p, 96, 97p	*Breakthroughs in Math / Bk. 1,* pages 18, 19, 44, 45, 72, 73, 100–101 *Number Sense,* Bk. 1, pages 10, 27, 30, 49 *Number Sense,* Bk. 2, pages 1–4, 21–24, 46

Number Theory **Pre-Test Score** ☐ **Post-Test Score** ☐

Subskill	TABE, Form 7 Pages	TABE, Form 8 Pages	Practice and Instruction Pages in This Text	Additional Practice and Instruction Resources
Odd and even numbers	40	29	6, 15p, 16p	
Sequence	30	4, 23, 45	88, 98p, 99p	
Properties	1		17, 18p, 31, 32p, 46, 47p, 58, 59p, 96, 97p	*Breakthroughs in Math / Bk. 1,* pages 18, 19, 44, 45, 72, 73, 100–101 *Foundations: Mathematics,* pages 14, 17, 30, 31, 39, 42, 44 *Number Sense,* Bk. 1, pages 10, 27, 30, 49
Equivalent form	26, 35	19, 43	7–10, 15p, 16p, 105	*Number Power,* Bk. 2, pages 5, 8–12 *Breakthroughs in Math / Bk. 2,* pages 71, 72 *Number Sense,* Bk. 5, pages 18–21 *Foundations: Mathematics,* pages 129–132
Multiples	6		88, 89	*Number Sense,* Bk. 1, page 13 *Critical Thinking with Math,* pages 4, 5

Data Interpretation **Pre-Test Score** ☐ **Post-Test Score** ☐

Subskill	TABE, Form 7 Pages	TABE, Form 8 Pages	Practice and Instruction Pages in This Text	Additional Practice and Instruction Resources
Graphs	32, 34	5, 7, 31, 32, 33	79–82, 84p, 85p	*Number Power,* Bk. 5, pages 6–66 *Breakthroughs in Math / Bk. 2,* pages 164–171 *The Math Problem Solver,* pages 218, 219, 223–225
Probability/ statistics		27	83, 84p, 85p	*Number Power,* Bk. 1, pages 126, 127 *Number Power,* Bk. 8, pages 60–64 *Breakthroughs in Math / Bk. 1,* pages 160, 161
Tables, charts, and diagrams	8, 9, 10, 11, 16	20, 21, 24, 25	73–78, 84p, 85p, 90p, 117	*Number Power,* Bk. 8, pages 13, 14, 34–37 *Breakthroughs in Math / Bk. 2,* pages 162–163 *Number Sense,* Bk. 1, pages 57, 60

Pre-Algebra **Pre-Test Score** ☐ **Post-Test Score** ☐

Subskill	TABE, Form 7 Pages	TABE, Form 8 Pages	Practice and Instruction Pages in This Text	Additional Practice and Instruction Resources
Function/pattern	13	42, 44	86–92, 98–100p	*Number Power Review,* pages 98, 99 *Critical Thinking with Math,* pages 3–5, 8–12
Missing element		1, 2, 3	91p, 92p, 96, 97p, 98–100p	*Number Power Review,* pages 5, 172, 173 *Number Sense,* Bk. 1, pages 18, 54, 55 *Foundations: Mathematics,* pages 15, 16, 18
Number sentence	12, 28, 37	15	93–95, 98, 100p	*The Math Problem Solver,* pages 2–6 *Critical Thinking with Math,* pages 16–19

Measurement **Pre-Test Score** ☐ **Post-Test Score** ☐

Subskill	TABE, Form 7 Pages	TABE, Form 8 Pages	Practice and Instruction Pages in This Text	Additional Practice and Instruction Resources
Appropriate instrument		12	101p, 110, 118p, 119p	
Calendar	45		77	*Real Numbers,* Bk. 4, pages 54, 62–64
Time	20, 24, 25	10	114–117, 118p, 119p	*Number Power,* Bk. 9, pages 128, 129, 131–135, 140, 141, 144 *Breakthroughs in Math / Bk. 1,* pages 140–153
Temperature	3	39	102, 106–109, 118p, 119p	*Number Power,* Bk. 9, pages 12, 13, 17, 74–77 *Real Numbers,* Bk. 4, pages 2–4, 7, 8, 46–52
Length	47	37, 38	102–108, 110, 111, 117, 118p, 119p	*Number Power,* Bk. 9, pages 10–13, 20–23, 25–27, 35–37 *Breakthroughs in Math / Bk. 2,* pages 149–152 *Number Sense,* Bk. 5, pages 27, 51–54

Geometry **Pre-Test Score** ☐ **Post-Test Score** ☐

Subskill	TABE, Form 7 Pages	TABE, Form 8 Pages	Practice and Instruction Pages in This Text	Additional Practice and Instruction Resources
Symmetry	49		125, 126–128p	*Real Numbers,* Bk. 6, 63
Pattern, shape	46	30, 48, 49	86–87, 98–100p	
Congruency	50		124, 126–128p	*Real Numbers,* Bk. 6, pages 60, 61
Plane figures			121, 126–128p	*Real Numbers,* Bk. 6, 16–20
Visualization	48	50	122, 132p, 126–128p	*Real Numbers,* Bk. 6, pages 25, 34, 35, 50–52
Logical Reasoning	7		120, 121, 126–128p	

Computation in Context **Pre-Test Score** ☐ **Post-Test Score** ☐

Subskill	**TABE, Form 7 Pages**	**TABE, Form 8 Pages**	**Practice and Instruction Pages in This Text**	**Additional Practice and Instruction Resources**
Whole Numbers	19, 22, 23, 33, 38, 41	16, 18	11, 12, 17, 26–30, 41–45, 53–57, 68–72p, 76, 80–85p, 93–95, 97–100p, 110p, 112–119p	*Number Power,* Bk. 1, pages 124, 125 *Breakthroughs in Math / Bk. 1,* pages 28–31, 38–41, 51–55, 65–69, 82, 83, 94–97, 113–115, 125–131, 134–139 *Number Sense,* Bk. 1, pages 25, 26, 46, 47, 50–54, 56–60
Decimals	14, 44	6, 9, 13, 14, 22, 26	11, 12, 26–30p, 41–45, 78	*Number Power,* Bk. 1, pages 119–123, 131 *Breakthroughs in Math / Bk. 1,* pages 14–16, 26–31, 38–41, 51–55, 65–69, 82, 83, 94–97, 113–115, 125–131, 134–135, 138 *Number Sense,* Bk. 3, pages 45–54, 56–59
Problem solving	39		26–30, 41–45, 53–57, 68–72p, 93–95p	*Number Power,* Bk. 6, pages 7–13, 27–35, 99–106 *Breakthroughs in Math / Bk. 1,* pages 38, 39, 65, 66, 125–127, 129–131, 134–139 *Number Sense,* Bk. 2, pages 19, 20, 47–60

Estimation **Pre-Test Score** ☐ **Post-Test Score** ☐

Subskill	**TABE, Form 7 Pages**	**TABE, Form 8 Pages**	**Practice and Instruction Pages in This Text**	**Additional Practice and Instruction Resources**
Reasonableness of Answer	18		12, 108, 112, 113	*Critical Thinking with Math,* pages 50, 51 *Real Numbers,* Bk. 1, pages 1, 2, 4, 16, 23, 30, 45, 50, 52, 59, 68 *Real Numbers,* Bk. 2, 1, 16, 23, 33, 37
Rounding	17, 21, 36, 42	8, 17, 46, 47	13, 14, 15p, 16p, 52	*Number Power Review,* pages 6, 7 *Breakthroughs in Math / Bk. 1,* pages 12, 13 *The Math Problem Solver,* pages 8, 56
Estimation	15, 29	11, 34, 35	11, 12, 25, 39, 40, 52, 67, 108	*Number Power,* Bk. 6, pages 36–41 *Number Power,* Bk. 8, pages 83–87 *Breakthroughs in Math / Bk. 1,* pages 92–95

The Number System

Place Value

The ten **digits** are **0, 1, 2, 3, 4, 5, 6, 7, 8,** and **9.** The value of a digit in a number depends on its location in that number. The locations, or **place values,** include ones, tens, hundreds, and thousands.

Look at these numbers:

53 35

They have the same digits, but they are different numbers. That's because the digits are in different places.

The number 53 stands for 50 + 3 or **5 tens and 3 ones.**
The number 35 stands for 30 + 5 or **3 tens and 5 ones.**

To understand numbers, you need to learn about place value.

These are the first four places in whole numbers.

thousands	hundreds	tens	ones
___	___	___	___

thousands	hundreds	tens	ones
___	___	8	9

thousands	hundreds	tens	ones
___	3	0	1

The number 89 has digits in two places.
It has 8 tens and 9 ones.

The number 301 has digits in three places.
It has 3 hundreds, 0 tens, and 1 one.

PRACTICE

Now you try it. Fill in these blanks.

1 953 has _______ hundreds, _______ tens, and _______ ones.

2 72 has 7 _______ and 2 _______.

3 30 has _______ tens and _______ ones.

4 45 has ____.

5 415 has ______________________.

6 201 has ______________________.

7 1,246 has ______________________.

8 What place value does the 6 have in the number 467? ______________

9 What place value does the 1 have in the number 1,042? ______________

10 What place value does the 3 have in the number 503? ______________

11 What place value does the 3 have in the number 329? ______________

12 What digit is in the hundreds place in the number 3,569? ______________

13 What digit is in the tens place in the number 1,479? ______________

14 What digit is in the hundreds place in the number 9,051? ______________

In the number 8,312, the 8 is in the thousands place. Its value is eight thousand or 8,000.

15 Tell the value of the 3 in 8,312.

____

16 Tell the value of the 1 in 8,312.

____

17 Tell the value of the 6 in 764.

____

18 Tell the value of the 4 in 6,421.

____

Naming Large Numbers

You can use the place value of each digit in a number to say the number aloud.

784 =	7 hundreds	+	8 tens	+	4 ones
=	700	+	80	+	4
=	seven hundred		eighty		four

The number written as 700 + 80 + 4 is in **expanded notation** form.

Pay close attention to the zeros in a number. If you see a zero in a place value, don't say that place value when you say the number.

2,509 = two thousand, five hundred nine	You do not say anything for the tens.
2,590 = two thousand, five hundred ninety	You do not say anything for the ones.
2,059 = two thousand, fifty-nine	You do not say anything for the hundreds.

PRACTICE

Write each number in expanded form.
(Example: 784 in expanded form is 700 + 80 + 4.)

1 319 ____________

2 1,649 ____________

3 2,061 ____________

Say each number, and write it in words.

4 546 ____________________________

5 601 ____________________________

6 4,130 ____________________________

7 1,041 ____________________________

8 3,009 ____________________________

Writing Large Numbers

Be careful when you use digits to write numbers from words. You have to use a zero if the place value is not mentioned. If you do not use zeros correctly, you may write some of the digits in the wrong place.

three hundred nine	=	300 + 9
	=	309 NOT 39

PRACTICE

Write each number in expanded form.
(Example: 970 in expanded form is 900 + 70 + 0.)

1 seven hundred six ______________________

2 one hundred forty ______________________

3 two thousand, thirteen ______________________

Write each number in digits.

4 three hundred two

____

5 one hundred sixty

____

6 one thousand, fifty

____

7 three thousand, six

____

8 four thousand, one hundred

____

9 nine thousand, one hundred

____

10 five thousand, three hundred six

____

11 six thousand, four hundred fifty

____

Comparing Numbers

These two whole numbers have the same number of places. Compare them by looking at the digits on the left.

3 7 8
5 0 3

The number 503 is larger than the number 378.

These two whole numbers have different numbers of digits. The number with more digits is larger.

2, 7 1 0
9 7 8

The number 2,710 is larger than the number 978.

These two whole numbers have the same number of digits.
Starting at the left, compare each pair of digits.

4, 1 8 8
4, 7 0 9

The number 4,709 is larger than the number 4,188.

PRACTICE

Circle the smaller number in each pair.

1 128 54

2 93 20

3 203 99

4 29 59

5 679 432

6 50 68

Circle the smallest number in each group.

7 51 92 13

8 79 97 53

9 160 98 29

10 13 99 54

Arrange these numbers from smallest to largest.

11 83, 105, 156, 12

____

12 100, 59, 7, 10

____

13 87, 75, 23, 65, 51

____

Arrange each set of digits to make the smallest number possible.

14 1, 8, 5

____

15 5, 9, 3

____

Special Types of Numbers

Here are three commonly used sets of numbers.

The **even numbers** can be divided evenly by two:
2, 4, 6, 8, 10, 12, 14, and so on.
Every even number ends in 0, 2, 4, 6, or 8.

The **odd numbers** cannot be evenly divided by two:
1, 3, 5, 7, 9, 11, 13, 15, and so on.
The last digit in any odd number is 1, 3, 5, 7, or 9.

First, second, third, fourth, and so on are called **ordinal numbers.** An ordinal number refers to a list. For most numbers ending in 1, 2, or 3, the ordinal number ends with *-first, -second,* or *-third.* (Examples: 41 becomes *forty-first* and 103 becomes *one hundred third.*) For all other numbers, including 11, 12, and 13, just write *-th* at the end of the number. (Examples: 38 becomes *thirty-eighth* and 212 becomes *two hundred twelfth.*)

PRACTICE

1 Circle each even number.

3, 6, 18, 15, 21, 32,
4, 5, 12, 48, 8, 103

2 Circle each odd number.

5, 8, 11, 7, 14, 21,
7, 32, 1, 45, 6, 13

Use this row of symbols to answer questions 3–6.
Start at the left to count the ordinal numbers

3 Name the third object in the row.

____

4 Which object is the letter?

____

5 Name the twelfth object in the row.

____

6 Which object is the airplane?

____

Fractions

A number such as $\frac{1}{2}$ (one-half) or $\frac{3}{4}$ (three-fourths) is a **fraction.** A fraction stands for **a part** of something.

Exactly $\frac{5}{8}$ of the pizza remains.

In a fraction, the bottom number tells how many parts are in the whole object.

The pizza was divided into 8 slices.
Each slice represents $\frac{1}{8}$ of the pizza.

PRACTICE

The top number in a fraction tells how many parts are shown.

In the pizza above, there are 5 slices remaining.

What fraction of each shape is shaded?

1 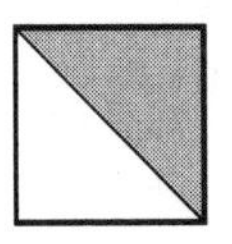

2 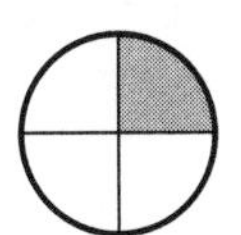

3 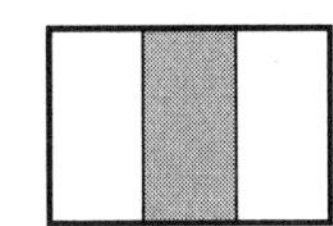

4 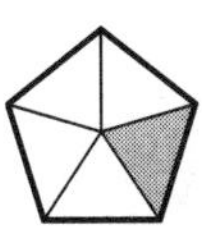

What fraction of each object is shaded?

5 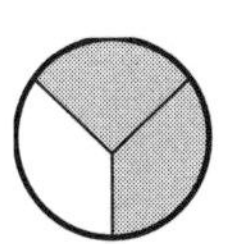

6 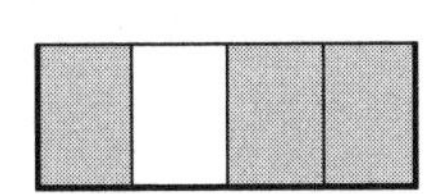

7 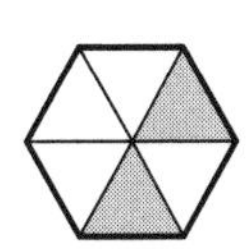

8 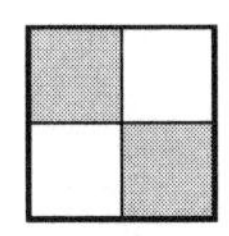

There is a fraction next to each rectangle. Using the bottom number of the fraction, divide the rectangle into parts. Using the top number, shade that number of parts.

9 $\frac{2}{3}$ 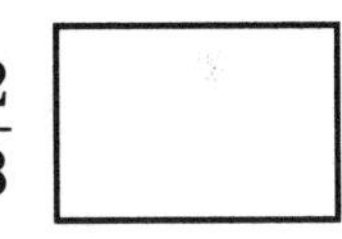

10 $\frac{1}{4}$ 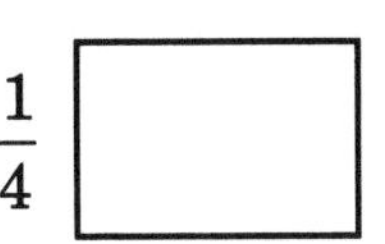

11 $\frac{2}{5}$ 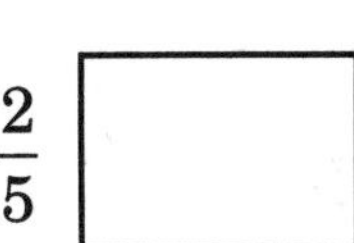

12 $\frac{3}{4}$ 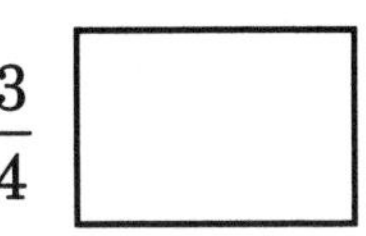

Mixed Numbers

A combination of a fraction and a whole number is called a **mixed number.**

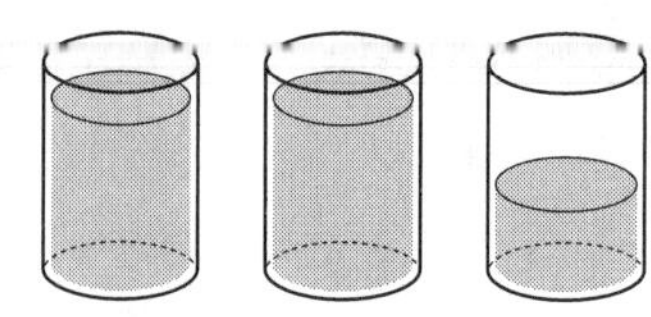

This is two and a half glasses of water.

$2 + \frac{1}{2}$ is written as $2\frac{1}{2}$.

$2\frac{1}{2}$ is a mixed number.

Write a mixed number for each problem.

1

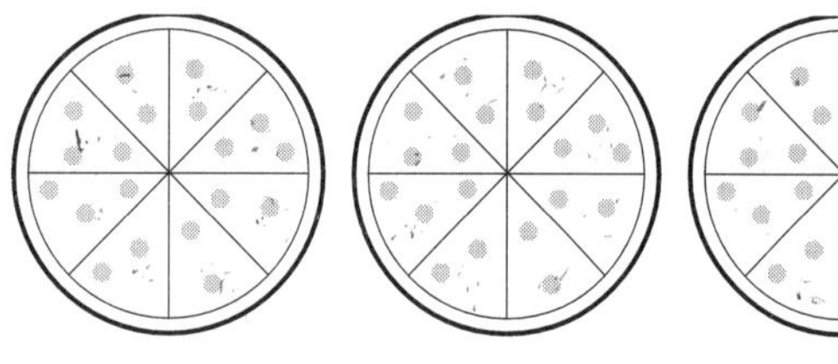

_____ pizzas

2

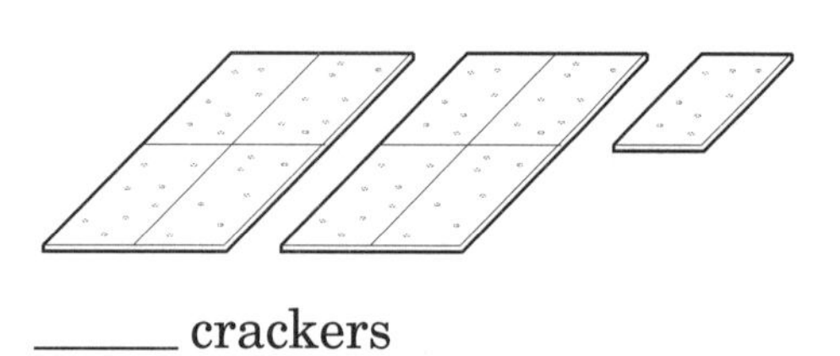

_____ crackers

3

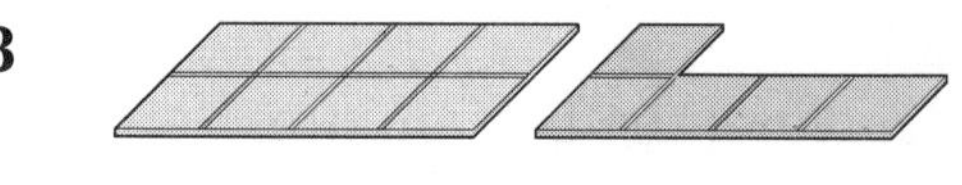

_____ candy bars

4

_____ cups

Write each of the fractions below. Remember, the bottom number is the number of parts in one whole object. The top number is the number of parts that are mentioned.

5 Two days is what fraction of a week? __________

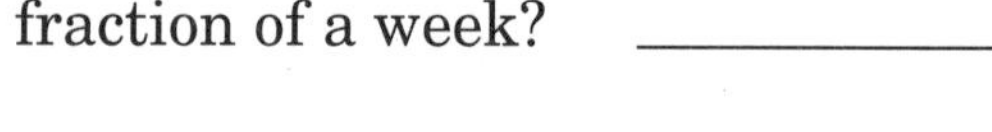

6 Five cents is what fraction of a dime? __________

7 There are 12 inches in a foot. What fraction of a foot is 5 inches?

____

8 There are nine workers in the office. Two of them will retire this year. What fraction of the workers will retire this year?

____

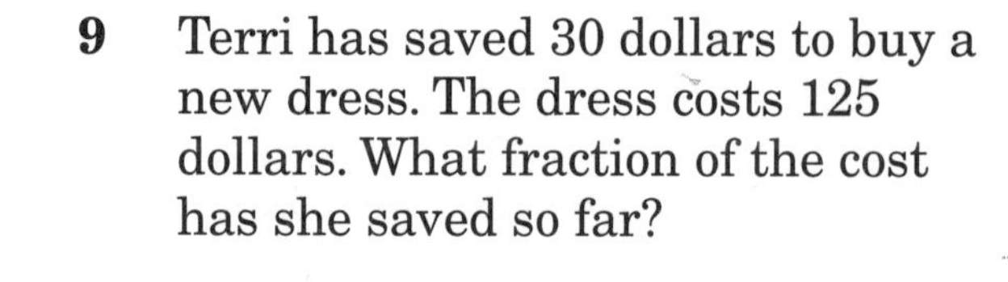

9 Terri has saved 30 dollars to buy a new dress. The dress costs 125 dollars. What fraction of the cost has she saved so far?

____

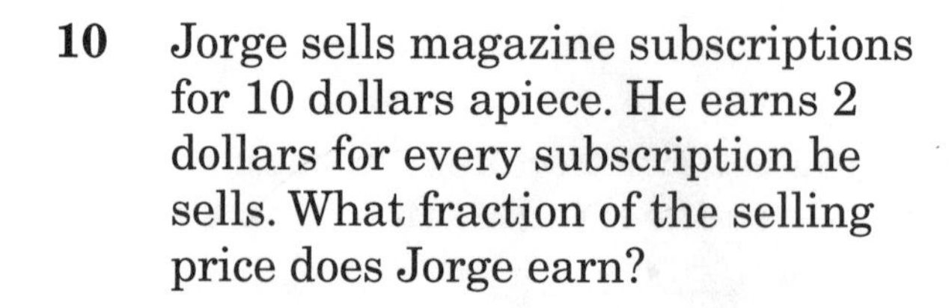

10 Jorge sells magazine subscriptions for 10 dollars apiece. He earns 2 dollars for every subscription he sells. What fraction of the selling price does Jorge earn?

____

11 Ramsey earns \$500 a week. He pays \$200 a week in taxes. What fraction of his weekly earnings does he pay in taxes?

____

Using Coins

These problems ask you to switch between one type of coin and another. Check your answers. If you get some wrong, erase your answers and try the problems again.

1 One nickel has the same value as _?_ pennies. ________

2 One dime has the same value as _?_ pennies. ________

3 One quarter has the same value as _?_ nickels. ________

4 One dime has the same value as _?_ nickels. ________

5 One quarter has the same value as _?_ dimes and _?_ nickels.

________ ________

penny 1¢

nickel 5¢

dime 10¢

quarter 25¢

Each coin represents a fraction of a dollar.

Each cent is $\frac{1}{100}$ of a dollar.

1 nickel $= \frac{5}{100}$ of a dollar

1 dime $= \frac{10}{100}$ of a dollar

1 quarter $= \frac{25}{100}$ of a dollar

Write each value as a fraction of a dollar.

6 8 cents is ________ of a dollar.

7 15 cents is ________ of a dollar.

8 50 cents is ________ of a dollar.

9 75 cents is ________ of a dollar.

10 Two dimes represent ________ of a dollar.

11 Five pennies represent ________ of a dollar.

12 Five dimes represent ________ of a dollar.

13 Two nickels represent ________ of a dollar.

Writing Money in Decimal Form

You can write amounts of money in **decimal form.** For example, you can write one dollar and five cents in decimal form as $1.05 Dollars are shown to the left of the **decimal point.** Cents are shown as two digits to the right of the decimal point. The word **and** represents the decimal point.

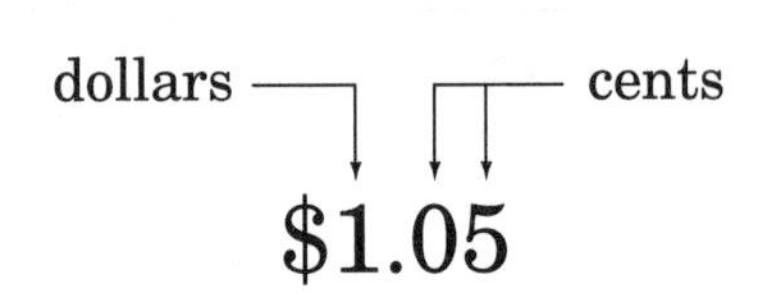

PRACTICE

Write each amount in words.
(Hint: $0.05 is *five cents,* $0.50 is *fifty cents,* and $9.15 is *nine dollars* ***and*** *15 cents.*)

1 $3.23 is ______________________________

2 $4.03 is ______________________________

3 $1.13 is ______________________________

4 $0.70 is ______________________________

When you write an amount of money in decimal form, always use two digits to the right of the decimal point.

Amount	Decimal Form
Five cents	$0.05
Two dollars and three cents	$2.03

Note: Never write "$1.5" because it is not clear whether that means *one dollar and five cents* or *one dollar and fifty cents.*

Write these amounts in decimal form.

5 four dollars and ten cents

____

6 three dollars and five cents

____

7 One dollar and fifty cents

____

8 Two dollars and six cents

____

9 Thirty-two cents

____

10 75 cents

____

11 3 cents

____

12 $\frac{1}{4}$ of a dollar

____

Estimation

Sometimes you need to know *about* how much something costs or *about* how long something takes. These amounts called **estimates.** An estimate is a number that is close to an actual amount.

PRACTICE

In each pair, circle the letter for the situation that calls for an estimate.

1 **A** Making change
  B Adding up the cost of groceries to make sure you have enough cash to pay

2 **F** Figuring out how much food you will need for a party
  G Adding up a bill for a customer

3 **A** Predicting how many people will attend a concert
  B Recording running times in a race

4 **F** Making out paychecks
  G Finding out how much gas you will need for a long trip

You can tell that a math problem calls for estimation because it uses signal words such as *almost, approximately,* and *about.*

5 **Circle the letter of each problem that calls for estimation. Do not try to solve the problems.**

A There are going to be 12 adults and 32 children at the party. About how many people will there be in all?

B What is the total cost of this meal?

Sandwich	$3.50
Drink	.75
Salad	1.50
subtotal	5.75
Tip	1.00
Total	____

C Soda costs 75 cents a can. About how many cans of soda can you buy for 5 dollars?

D There are 97 rows in the theater. Each row seats 17 people. Approximately how many people does the theater hold?

E Enrique drove 315 miles at 65 miles an hour. About how many hours did the trip take?

Common Sense Estimation

You often can use common sense or general information to make an estimate. For instance, you know from experience that a hamburger and fries should cost about five dollars, not fifty cents and not fifty dollars. This type of common sense estimation is very useful when you are doing math problems. *Get into the habit of using it to check whether an answer makes sense.*

For each question, circle the answer that makes the most sense.

1 You buy 12 gallons of gas. About how much should it cost?

A $0.80
B $18.00
C $80.00
D $180.00

2 If you are filling out a 5-page application and it takes 10 minutes to complete each page, about how long should it take to complete the entire application?

F 5 hours
G 15 minutes
H 1 hour
J 24 hours

3 Ryan buys a shirt for $14.95. He pays for it with a 20-dollar bill. His change should be just over which amount?

A 10 cents
B 50 cents
C 1 dollar
D 5 dollars

4 You buy a new washing machine for $\frac{1}{4}$ off. About how much will you save?

F 4 dollars
G 12 dollars
H 80 dollars
J 600 dollars

Circle the letter of the best answer to each question.

5 Which statement is not reasonable?

A Two large pizzas will feed about 200 children.
B It should take about 6 hours to drive 300 miles.
C To make one sandwich, you need about 2 slices of bread.
D One person can spend about 50 dollars a week on groceries.

6 Which statement is not reasonable?

F Students learn best when there is one teacher for every ten learners.
G If it takes 5 days for one person to finish a quilt, three people can finish one in an hour.
H It will cost about $15.00 to buy matinee movie tickets for a family of four.
J You should be able to drive about 180 miles on 10 gallons of gas.

Rounding

You can estimate the answer to a math problem by using **rounded** numbers.

Examples of Rounding

Exact Value		Rounded Value
32	workers	about 30 workers
52	weeks	a little over 50 weeks
$28\frac{1}{2}$	hours	almost 30 hours
193	students	about 200 students
$17.50		nearly $18.00

To round a number, think of the number as being part of a hilly number line like the one at the right.

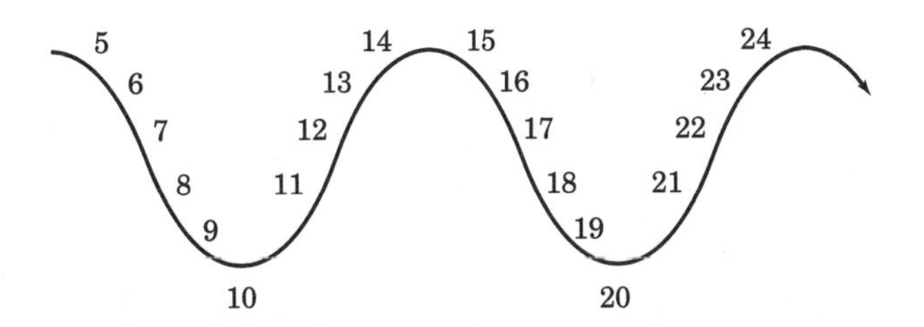

Every low spot ends in a zero. Numbers ending in 0, 1, 2, 3, and 4 roll **back** to the closest low spot. Numbers ending in 5, 6, 7, 8, and 9 roll **ahead** to the closest low spot.

Examples:

11 rounds to 10.
23 rounds to 20.
18 rounds to 20.
5 rounds to 10.

PRACTICE

Circle the correct answer.

1	21 rounds to	20	30
2	17¢ rounds to	10¢	20¢
3	53 rounds to	50	60
4	7/8 rounds to	0	1
5	113 rounds to	110	120
6	$1.03 rounds to	$0.90	$1.00
7	$1\frac{1}{4}$ rounds to	1	2

Round each number to the nearest ten.

8 8 rounds to ______.

9 12 rounds to ______.

10 21 rounds to ______.

11 142 rounds to ______.

12 57 cents rounds to ______.

13 97 cents rounds to ______.

14 118 rounds to ______.

You can round a number to different place values. For example:

To the nearest ten, **192** rounds to **190.**
To the nearest hundred, **192** rounds to **200.**

When you round to the **tens place,** all digits to the right of the tens place become zero. When you round to the **hundreds place,** all digits to the right of the hundreds place become zero.

> To round a number to a place value, look at the digit just to the right of that place value. If that digit is less than 5, round down. If that digit is 5 or more, round up.

Examples:
To round 482 to the nearest ten, look at the digit to the right of the tens place. It is 2, so 482 rounds to 480.
To round 482 to the nearest hundred, look at the digit to the right of the hundreds place. It is 8, so 482 rounds to 500.

PRACTICE

Round each number to the indicated place.

15 Round 145 to the tens place. ________.

16 Round 281 to the tens place. ________.

17 Round 453 to the hundreds place. ________.

18 Round 671 to the hundreds place. ________.

19 Round 1,204 to the thousands place. ________.

20 Round 2,456 to the thousands place. ________.

21 Round 345 to the tens place. ________.

22 Round 567 to the hundreds place. ________.

23 Round 1,340 to the hundreds place. ________.

Number Systems Skills Practice

Circle the letter for the correct answer to each problem.

1 Which group of numbers is in order from smallest to largest?

A 34, 8, 41, 46, 19
B 8, 19, 34, 41, 46
C 19, 34, 41, 46, 8
D 8, 19, 34, 46, 41

2 Which of these numbers is another name for three hundreds, zero tens, six ones?

F 36
G 316
H 360
J 306

3 What is 349 rounded to the nearest ten?

A 340
B 400
C 300
D 350

4 A student has a five-dollar bill and three pennies. Which of these numbers shows how much money she has?

F \$5.3 **H** \$5.03
G \$5.30 **J** \$0.53

5 Ellen's new boss says she will earn two thousand, thirty dollars each month. Which of these numbers should be on her monthly paycheck?

A \$2,030 **C** \$2,300
B \$230 **D** \$23,000

6 Lionel needs an even number of children in his dance class. Which of these is an even number?

F 20 **H** 15
G 9 **J** 17

7 Carlos buys a hamburger for \$3.23, fries for \$0.49, and a soda for \$0.79. Which of these is the most reasonable estimate of how much his lunch cost?

A \$20.00
B \$5.00
C \$2.00
D \$15.00

8 Gina had a 20-dollar bill. Then she spent 5 dollars. What fraction of the money did Gina spend?

F $\frac{20}{5}$ **H** $\frac{1}{20}$

G $\frac{1}{5}$ **J** $\frac{5}{20}$

A lot of cars pass Marisa's house during rush hour. Marisa thinks a stop sign would make her corner more safe. She wants to prove this to the city. So, every day between 5:00 and 7:00 she counts the cars that pass by. This list shows how many cars she counted last week. **Use the list to answer questions 9 through 12.**

Monday	200
Tuesday	196
Wednesday	231
Thursday	213
Friday	230
Total	**1,070**

9 On which day did the most cars pass?

A Monday
B Tuesday
C Wednesday
D Thursday

10 How many cars passed by on the fourth day?

F 200
G 196
H 231
J 213

11 When Marisa reports the total number of cars that passed, what should she say?

A one hundred seven
B one hundred seventy
C one thousand, seven hundred
D one thousand, seventy

12 What fraction of the total cars passed by on Monday?

F $\frac{200}{1070}$ **H** $\frac{200}{196}$

G $\frac{1070}{200}$ **J** $\frac{870}{200}$

13 Which group of values is in order from smallest to largest?

A \$1.95, \$2.00, \$1.59, \$0.79
B \$1.95, \$1.59, \$2.00, \$0.79
C \$1.59, \$1.95, \$2.00, \$0.79
D \$0.79, \$1.59, \$1.95, \$2.00

14 Which group of coins is worth $\frac{1}{2}$ of a dollar?

F 3 dimes and 1 nickel
G 3 dimes and 1 quarter
H 3 nickels and 1 dime
J 2 nickels and 4 dimes

Addition

Basic Concepts

When you put things together, the math term is **addition** or **adding.**

- Some of the terms that signal "addition" are *plus, sum, total, added to, altogether,* and *in all.* Here are some examples:
  five *plus* three
  the *sum* of seven and two
  the *total* of one and three
  six *added to* three
  how many *altogether*
  how many *in all*

- In a math sentence, the symbol "+" tells you to add.

  The result of adding is called a **sum** or a **total.**

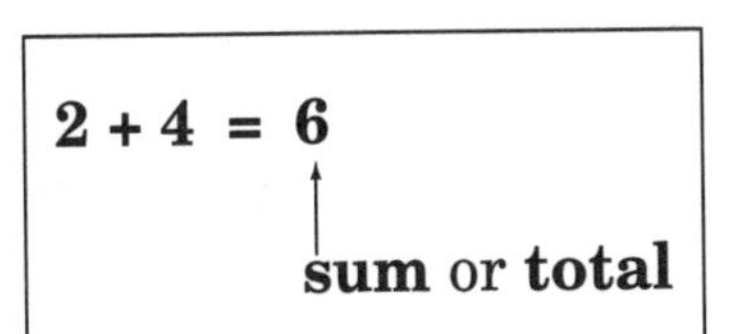

- An addition problem can be written from left to right in a **row,** or from top to bottom in a **column.**

in a row *or* **in a column**

2 + 4 = 6 (↑ **sum**)

$$\begin{array}{r} 2 \\ +\,4 \\ \hline 6 \end{array}$$
sum → 6

- You can change the order of the numbers when you add. That does not change the answer.

The value of **2 + 4** is the same as **4 + 2.**

- If you add zero to a number, you do not change the value.

2 + 0 = 2

- You can add ***like*** things together. You can add 4 hours and 3 hours, but you *cannot* add 3 hours and 5 miles. You *cannot* add 3 hours and 65 minutes, but you *can* add 180 minutes and 65 minutes.

PRACTICE

1 **Circle the letter for each word problem that calls for adding. *Do not try to solve the problems.***

A What is the sum of 3 and 2?

B There are 15 oaks and 3 maples in the park. What is the total number of trees?

C There are 6 women and 4 men on a work crew. How many workers are there altogether?

D Eric grew 46 pumpkins and sold 30. How many pumpkins are left?

E What is 31 apples minus 10 apples?

F What is 15 plus 9?

G Tina spent 6 dollars at the book store and spent 12 dollars on gas. How much did she spend in all?

H Reginald's grocery bill was 15 dollars. He paid with a 20-dollar bill. How much change will he get back?

I Jorge has to serve one pizza fairly to 4 children. How much of the pizza does each child get?

Write T for *true* or F for *false*.

2 The value of 41 + 13 is the same as 13 + 41. ____________

3 The value of 2 + 5 is the same as 7 + 2. ____________

4 4 + 0 = 40 ____________

5 The value of 2 + 5 is the same as 2 – 5. ____________

6 The value of 62 + 0 is the same as 62. ____________

7 The value of 1 + 3 is the same as 3 + 1. ____________

Basic Addition Facts

To solve problems using addition, you will need to add small numbers very quickly—almost without thinking. This page will give you practice.

PRACTICE

Fill in all the blanks on this page. If you make more than a few mistakes, erase your answers and try the problems again.

1	2 + 3 = ______	3 + 3 = ______	0 + 1 = ______	7 + 7 = ______
2	8 + 8 = ______	6 + 4 = ______	2 + 2 = ______	9 + 6 = ______
3	1 + 1 = ______	6 + 5 = ______	6 + 3 = ______	1 + 3 = ______
4	5 + 2 = ______	4 + 7 = ______	0 + 0 = ______	4 + 1 = ______

5	4 + 3 ___	7 + 8 ___	4 + 4 ___	8 + 9 ___	9 + 9 ___	4 + 0 ___
6	7 + 0 ___	7 + 5 ___	3 + 5 ___	2 + 4 ___	1 + 2 ___	3 + 0 ___
7	5 + 1 ___	6 + 2 ___	1 + 7 ___	8 + 4 ___	6 + 6 ___	3 + 7 ___
8	8 + 2 ___	3 + 8 ___	0 + 2 ___	1 + 8 ___	5 + 4 ___	0 + 5 ___
9	9 + 4 ___	5 + 5 = ______	8 + 7 ___	6 + 1 = ______	8 + 0 ___	
10	8 + 5 ___	9 + 3 = ______	7 + 9 ___	0 + 6 = ______	5 + 8 ___	

Adding Three or More Numbers

You always add two numbers at a time—even when you have three or more numbers to add. To add several numbers together, find the sum of the first two numbers. Then add the next number to that sum. Keep adding the "next" number until you complete them all.

2 ⎤	
3 ⎦	→ *Add the first two numbers.*
2	*Then, add the next one:* 5 + 2 = 7
+ 4	*Finally, add the last number:* 7 + 4 = 11
11	*The sum is 11.*

PRACTICE

Solve each problem. Then add the numbers in a different order to check your answers.

1

$$\begin{array}{r}1\\2\\+\,3\\\hline\end{array}\quad\begin{array}{r}3\\2\\+\,8\\\hline\end{array}\quad\begin{array}{r}2\\1\\+\,4\\\hline\end{array}\quad\begin{array}{r}4\\2\\+\,9\\\hline\end{array}\quad\begin{array}{r}6\\2\\+\,9\\\hline\end{array}\quad\begin{array}{r}2\\2\\+\,0\\\hline\end{array}$$

2

$$\begin{array}{r}3\\2\\+\,4\\\hline\end{array}\quad\begin{array}{r}5\\1\\+\,3\\\hline\end{array}\quad\begin{array}{r}4\\2\\+\,3\\\hline\end{array}\quad\begin{array}{r}8\\1\\+\,1\\\hline\end{array}\quad\begin{array}{r}3\\4\\+\,1\\\hline\end{array}\quad\begin{array}{r}3\\4\\+\,3\\\hline\end{array}$$

3 4 + 4 + 4 = ______ 5 + 0 + 2 = ______ 6 + 1 + 2 = ______

4 5 + 5 + 8 = ______ 7 + 2 + 2 = ______ 9 + 1 + 6 = ______

5 3 + 5 + 2 = ______ 9 + 0 + 6 = ______ 6 + 2 + 8 = ______

Adding in Column Form

To set up an addition problem in column form, write one number above the other. Be sure the columns line up. It doesn't matter which number is on top.

Start at the right. Add the digits in that column. Then move one column to the left, and add the digits in that column. Keep this up until you have added all the columns.

You can check your work. One way is to add each column again, from the bottom to the top.

PRACTICE

Find each sum. Then check your answer by adding the columns from the bottom to the top.

1	41 + 13	220 + 361	21 + 52	126 + 113	47 + 41
2	15 + 80	37 + 11	353 + 326	66 + 32	74 + 15
3	761 + 213	68 + 31	32 + 42	206 + 403	12 + 17

If the top number in a problem has a dollar sign or a decimal, such as $1.20 shown at the right, repeat them in the answer. Decimal points and labels line up vertically.

$1.20 1.23 $2.43

4	$61 + 30	$2.08 + 3.00	22 inches + 31 inches	$1.20 + 3.40	14 miles + 54 miles

Adding Small and Large Numbers

To add numbers like 4 and 23, **line up the digits in the ones column.**

You can think of the empty space as a zero. Start at the right, and add the columns.

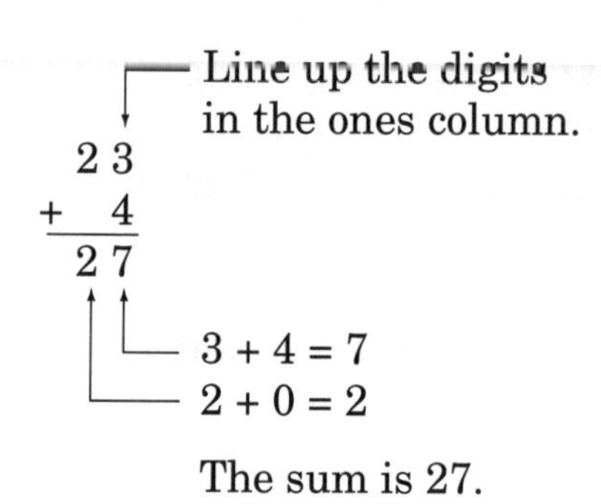

The sum is 27.

PRACTICE

Find each sum. Then check your answers. You can use common-sense estimation or bottom-to-top addition.

1	71 + 3	514 + 32	38 + 160	94 + 5	26 feet + 2 feet
2	48 + 1	602 + 26	80 + 7	19 + 310	$77 + 2

Rewrite each addition problem in column form. Be sure to line up the digits in the ones place. Find each sum and check your answers.

3 13 + 4 = ______

4 6 + 23 = ______

5 $10 + $9 = ______

6 501 + 48 = ______

7 60 + 301 = ______

8 5 + 931 = ______

When a Column Sum Is More Than 9

In this addition problem, the sum of the right column is the 2-digit number 14. Write the 4 in the ones place. Write the other digit, 1, at the top of the next column. Then add all three digits in the tens column.

$$\begin{array}{r} {}^{1}47 \\ +\ \ 7 \\ \hline 54 \end{array}$$

7 + 7 = 14

Write the 4. Write 1 above the tens place.

1 + 4 = 5

When a column sum is more than 9, write the digit in the ones place below the column. Write the other digit or digits at the top of the next column. Then continue to add.

$$\begin{array}{r} {}^{1}463 \\ +172 \\ \hline 635 \end{array}$$

6 + 7 = 13
Write 3 in the tens column.
Write 1 above the hundreds column.

PRACTICE

Find each sum. Check your answers.

1	48 + 4	96 + 12	85 + 7	19 + 3	77 + 4
2	27 + 4	45 + 15	58 + 60	49 + 5	$26 + 6
3	48 + 103	602 + 9	90 + 57	9 inches + 31 inches	77 + 3
4	206 + 6	55 + 5	39 + 52	19 + 9	464 + 50

5	81 + 23	106 + 9	$71.00 + 19.00	19 + 7	99 + 10
6	44 + 7	75 + 25	83 + 7	93 + 24	19 cups + 8 cups

Rewrite each problem in column form. Be sure to line up the digits in the ones place. Find each sum and check your answers.

7 19 + 4 = _______

8 63 + 7 = _______

9 9 + 32 = _______

10 30 + 80 = _______

11 82 + 80 = _______

12 26 + 19 = _______

13 107 + 15 = _______

14 6 + 29 = _______

15 50 + 651 = _______

16 507 + 107 = _______

17 10 + 90 = _______

18 51 + 80 = _______

19 330 + 99 = _______

20 25 + 80 = _______

Using Estimation To Check Addition

You can check a sum by using common sense or by adding columns from bottom to top. Another way to check addition uses the first digit in each number. This method is called **front-end** estimation.

Add the digits in the left column of each number. Write a zero for each of the remaining columns

Add the digits in the left column.

```
  3 6 1
+ 6 5 2
  9 0 0
```

Write zeros in all the other columns.

With this method, your estimate will always be *less* than the actual sum.

PRACTICE

Estimate each sum using the first digit of each number and front-end estimation. You do not have to find the exact answers. *Reminder:* Your estimate should have at least as many places as the largest number in the problem.

1   32 + 43   27 + 31   51 yards + 3 yards $97.00 + 2.00

2   217 + 433   302 + 833   372 + 413   527 + 33

Circle the letter for the *best* estimate to each problem.

3   65 + 31

A 30
B exactly 90
C a little less than 90
D a little more than 90

4 $3 + $52 = ____

F a little more than $3
G a little more than $8
H a little more than $50
J a little more than $80

5   465 + 69 = ____

A 46
B 1,000
C a little less than 400
D more than 400

6   605 + 14

F exactly 600
G exactly 700
H a little more than 600
J a little less than 700

Solving Word Problems

Here are five steps to use when solving a word problem.

1. Identify the particular question in the problem.
2. Get all the information you need.
3. Set up the problem.
4. Solve it.
5. Check your work.

You have had practice with the first step. Remember that the signal words *plus, sum, total, added to, altogether,* and *in all* usually call for addition.

PRACTICE

1 Circle the letter of each problem that calls for addition. *Do not try to solve the problems.*

A Erlene's son watched 8 hours of television on Saturday and 5 hours on Sunday. (He is supposed to watch no more than 2 hours a day.) How much television did he watch in all?

B You have a board that is 16 inches long. You cut off a piece that is 5 inches long. How much of the board is left?

C There are 16 men and 14 women in a math study group. If 2 more students join the group, how many will there be altogether?

D Julio earns $700 a week. Exactly $150 is taken out of his weekly paycheck for federal taxes and $70 is taken out for city and state taxes. How much tax is taken out of his paycheck each week?

E Last fall, Jason's daughter was 43 inches tall. This fall she is 47 inches tall. How much has she grown?

F Bree made 14 fruit cakes for the holidays. She gave 7 cakes to her cousins. How many fruit cakes were left?

G The gas tax in one state is 8 cents a gallon. In a second state, the gas tax is 15 cents higher. What is the gas tax in the second state?

H Ross works for a moving company. Last week he packed 125 boxes. This week he packed 154 boxes. How many more boxes did he pack this week?

I Kaitlyn supervises three men and four women out of the 32 people in her department. How many people does she supervise?

In each problem, circle the numbers you need to solve that problem. Then set up the problem. *Remember:* You can only add like things.

Sample. Erlene's son watched ⑧ hours of television on Saturday and ⑤ hours on Sunday. (He is supposed to watch no more than 2 hours a day.) How much television did he watch in all?

$$\begin{array}{r} 8 \\ +5 \\ \hline 13 \end{array}$$

2 There are 16 men and 14 women in a math study group. If 2 more students join the group, how many will there be altogether?

3 Julio earns $700 a week. Exactly $150 is taken out of his weekly paycheck for federal taxes and $70 is taken out for city and state taxes. How much is taken out of his paycheck each week?

4 The gas tax in one state is 8 cents a gallon. In a second state, the gas tax is 15 cents higher. What is the gas tax in the second state?

5 Kaitlyn supervises three men and four women out of the 32 people in her department. How many people does she supervise?

6 Amy and Eddie invited 87 people to their wedding. Soon, 54 people sent notes saying they would attend and 21 people sent notes saying they could not attend. Another 12 people called Amy or Eddie and said they would be there. How many guests will attend the wedding?

In many problems, you have to find out some information before you can solve the problem. On each blank line, tell what information is needed to solve the problem.

7 Mrs. Wu spent $7.00 at the dry cleaners. Then she bought 3 bags of groceries. How much did she spend in all?

You also need to know

____

____

8 It takes 12 hours to drive from Cincinnati to Broken Bow. If Marcus leaves Cincinnati now, what time will he arrive?

You also need to know

____

____

9 The shop can fix Mr. Patel's car for $210.00 plus the cost of parts. How much will his final bill be?

You also need to know

____

____

10 At his garage sale, Anthony sold albums for $1.25 each. How much money did he make selling albums?

You also need to know

____

____

Follow the five-step process on page 26 to solve these problems. Be sure to show all your work and show how you checked your answers.

11 There are five people in Freddy's family. She invited another thirteen people to Thanksgiving dinner. How many people will be at the dinner?

12 Ernesto usually works just eight hours a day. Saturday he worked eleven hours. Sunday he worked nine hours. How many hours did he work over the weekend?

Addition Skills Practice

Circle the letter for the correct answer to each problem.
Try crossing out all unreasonable answers before you begin to work.

1

$$\begin{array}{r} 36 \\ +\ 51 \\ \hline \end{array}$$

A 25
B 87
C 85
D 97
E None of these

2

$$\begin{array}{r} 700 \\ +\ 700 \\ \hline \end{array}$$

F 140
G 170
H 1400
J 1700
K None of these

3

$$\begin{array}{r} 27 \\ +\ 16 \\ \hline \end{array}$$

A 11
B 33
C 41
D 43
E None of these

4

42 + 56 = ____

F 98
G 18
H 94
J 14
K None of these

5

$$\begin{array}{r} 80 \\ +\ 60 \\ \hline \end{array}$$

A 140
B 180
C 120
D 1400
E None of these

6

22 + 88 = ____

F 100
G 106
H 102
J 110
K None of these

Use this information for questions 7 through 9.
Sandra works as a clerk. She earns $5.25 per hour. On holidays she gets twice that amount.

7 How much does Sandra earn per hour on holidays?

A $2.50
B $5.00
C $10.50
D $12.00

8 Sandra worked 37 hours last week and 42 hours the week before. How can you find out how many hours she worked in all?

F Add.
G Divide.
H Multiply.
J Subtract.

9 Sandra's boss gives her a raise of 65¢ per hour. How much is she now paid per hour for work on regular days?

A $4.60
B $5.65
C $5.85
D $5.90

10

$$\begin{array}{r} 403 \\ +\ 251 \\ \hline \end{array}$$

F 653
G 754
H 681
J 554
K None of these

11

$$\begin{array}{r} 2 \\ 4 \\ +\ 1 \\ \hline \end{array}$$

A 6
B 5
C 7
D 9
E None of these

12 13 + 25 = ____

F 38
G 28
H 37
J 48
K None of these

13

$$\begin{array}{r} 65 \\ +\ 25 \\ \hline \end{array}$$

A 80
B 90
C 800
D 900
E None of these

14 50¢ + 70¢ = ____

F 12¢
G $12.00
H $1.20
J $1.02
K None of these

15 156 + 9 = ____

A 155
B 167
C 166
D 165
E None of these

16

$$\begin{array}{r} 337 \\ 216 \\ +\ 200 \\ \hline \end{array}$$

F 553
G 755
H 753
J 743
K None of these

17 Which of these, if any, has the same value as 3 + 6 ?

A the difference between three and six
B the total of three and six
C the product of three and six
D one-third of six
E None of these

18 Which of these, if any, does <u>not</u> have the same value as 10 + 14?

F 14 + 10
G

$$\begin{array}{r} 10 \\ +\ 14 \\ \hline \end{array}$$

H 10 + 14 + 0
J 1 + 10 + 14
K All of them have the same value.

19 Which of these, if any, does <u>not</u> equal 9?

A 9 + 0
B 4 + 5
C 8 + 1
D

$$\begin{array}{r} 3 \\ +\ 6 \\ \hline \end{array}$$

E Each of them is equal to 9.

Subtraction

Basic Concepts

When you take objects away from a group, the math term is **subtraction** or **subtracting.**

- Some of the terms that signal "subtract" are *minus, take away, left over, difference, how much change,* and *how much more or less.* Here are some examples:

  What is the *difference* of 10 and 3?
  How much is *left over* when 12 is *taken away* from 20?
  What is the *change* from 4 to 9?

- To subtract, you need like things. You cannot subtract 4 apples from 10 oranges.

- You can write a subtraction problem in a row from left to right, or in columns from top to bottom. The symbol "–" is the subtraction symbol. The result of a subtraction problem is called the ***difference.***

  In a column: $\begin{array}{r} 5 \\ -3 \\ \hline 2 \end{array}$

  In a row: → $5 - 3 = 2$

  Difference → 2

- If you subtract zero from a number, you do not change the value: $2 - 0 = 2$

- If you subtract a number from itself, the result or difference is zero: $7 - 7 = 0$

- The order of the numbers in a subtraction problem is important: **4 – 2** and **2 – 4** have different values.

- Subtraction and addition are the opposite of each other. To reverse the result of a subtraction problem, add the same number.

Subtract 12:	Add 12:
$42 - 12 = 30$	$30 + 12 = 42$

  Also, to reverse the result of an addition problem, subtract the same number.

Add 17:	Subtract 17:
$53 + 17 = 70$	$70 - 17 = 53$

- You can subtract only one number at a time. In any problem, always start with the two numbers at the left.

  $$\begin{aligned} 9 - 3 - 2 &= (9 - 3) - 2 \\ &= (6) - 2 \\ &= 4 \end{aligned}$$

PRACTICE

1 Circle the letter of each word problem that calls for subtraction. You do not have to solve the problems.

A Isabella paid \$8.99 for a book, plus 72¢ tax. How much did she pay in all?

B One can of peas costs 69¢. Another costs 82¢. What is the difference between the two prices?

C DeShawn made 25 belt buckles. He sold 15. How many did he have left?

D Reuben spent 3 dollars for a hot dog. He paid with a 5-dollar bill. How much change should he get back?

E Luis earned \$120 for repairing a porch. He worked at the job for 5 hours. How much money did he earn per hour?

F What is 31 hours minus 10 hours?

Fill in the blanks.

2 If 82 + 61 = 143, then 143 − ______ = 82.

3 If 117 − 56 = 61, then 61 + ______ = 117.

4 If 75 + 32 = 107, then ________________________________.
(Write a subtraction problem.)

5 If 189 − 56 = 133, then ________________________________.
(Write an addition problem.)

6 9 − 4 − 2 = ______ − ______

7 10 − 5 − 3 = ______ − ______

Write T for *true* or F for *false*.

8 42 − 31 has the same value as 31 − 42. ______

9 12 − 5 has the same value as 12 + 5. ______

10 43 − 0 = 43 ______

11 142 − 142 = 0 ______

12 8 − 4 − 2 has the same value as 4 − 2. ______

13 7 − 4 − 1 has the same value as 7 − 3. ______

14 If 982 − 159 = 823, then 823 + 159 = 982. ______

Basic Subtraction Facts

If **2** + **3** = **5,** then **5** − **3** = **2** and **5** − **2** = **3.**

For most basic addition facts, there are two related basic subtraction facts. Find each difference below. If you make more than a few mistakes, try them again. You should be able to answer any basic-fact problem quickly and accurately.

PRACTICE

1 $8 - 6 =$ ______ $7 - 3 =$ ______ $1 - 0 =$ ______ $10 - 5 =$ ______

2 $8 - 8 =$ ______ $6 - 4 =$ ______ $10 - 2 =$ ______ $9 - 6 =$ ______

3 $10 - 1 =$ ______ $6 - 5 =$ ______ $6 - 3 =$ ______ $3 - 1 =$ ______

4 $5 - 2 =$ ______ $7 - 4 =$ ______ $3 - 2 =$ ______ $10 - 4 =$ ______

5 $\begin{array}{r} 4 \\ -\ 3 \\ \hline \end{array}$ $\begin{array}{r} 8 \\ -\ 7 \\ \hline \end{array}$ $\begin{array}{r} 4 \\ -\ 4 \\ \hline \end{array}$ $\begin{array}{r} 9 \\ -\ 8 \\ \hline \end{array}$ $\begin{array}{r} 9 \\ -\ 3 \\ \hline \end{array}$ $\begin{array}{r} 4 \\ -\ 0 \\ \hline \end{array}$

6 $\begin{array}{r} 7 \\ -\ 0 \\ \hline \end{array}$ $\begin{array}{r} 7 \\ -\ 5 \\ \hline \end{array}$ $\begin{array}{r} 5 \\ -\ 3 \\ \hline \end{array}$ $\begin{array}{r} 4 \\ -\ 2 \\ \hline \end{array}$ $\begin{array}{r} 2 \\ -\ 1 \\ \hline \end{array}$ $\begin{array}{r} 3 \\ -\ 0 \\ \hline \end{array}$

7 $\begin{array}{r} 5 \\ -\ 1 \\ \hline \end{array}$ $\begin{array}{r} 6 \\ -\ 2 \\ \hline \end{array}$ $\begin{array}{r} 7 \\ -\ 1 \\ \hline \end{array}$ $\begin{array}{r} 8 \\ -\ 4 \\ \hline \end{array}$ $\begin{array}{r} 7 \\ -\ 6 \\ \hline \end{array}$ $\begin{array}{r} 10 \\ -\ 3 \\ \hline \end{array}$

8 $\begin{array}{r} 8 \\ -\ 2 \\ \hline \end{array}$ $\begin{array}{r} 8 \\ -\ 3 \\ \hline \end{array}$ $\begin{array}{r} 10 \\ -\ 2 \\ \hline \end{array}$ $\begin{array}{r} 8 \\ -\ 1 \\ \hline \end{array}$ $\begin{array}{r} 5 \\ -\ 4 \\ \hline \end{array}$ $\begin{array}{r} 5 \\ -\ 0 \\ \hline \end{array}$

9 $\begin{array}{r} 9 \\ -\ 4 \\ \hline \end{array}$ $5 - 5 =$ _____ $\begin{array}{r} 10 \\ -\ 7 \\ \hline \end{array}$ $6 - 1 =$ _____ $\begin{array}{r} 8 \\ -\ 0 \\ \hline \end{array}$

10 $\begin{array}{r} 8 \\ -\ 5 \\ \hline \end{array}$ $9 - 3 =$ _____ $\begin{array}{r} 9 \\ -\ 7 \\ \hline \end{array}$ $10 - 6 =$ _____ $\begin{array}{r} 9 \\ -\ 5 \\ \hline \end{array}$

Subtracting in Column Form

To set up a subtraction problem in column form, write one number above the other. **The number you are subtracting *from* must be on top.**

Start at the right. Subtract the digits in that column. Then move to the left, and subtract the digits in that column. Continue until you have subtracted all the columns.

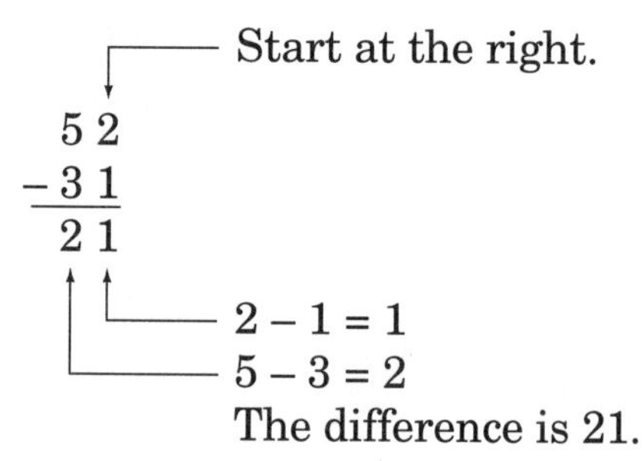

You can check your work. One way is to add the answer and the bottom number. The sum should be the same as the top number.

$$\begin{array}{r} 31 \\ +\,21 \\ \hline 52 \end{array}$$

Add the second number and the difference. (31 + 21)

That is the same as the original top number. (52)

PRACTICE

Find each difference. Then use addition to check your answer.

1 $\begin{array}{r} 76 \\ -\ 23 \\ \hline \end{array}$ $\begin{array}{r} 342 \\ -\ 220 \\ \hline \end{array}$ $\begin{array}{r} 521 \\ -\ 200 \\ \hline \end{array}$ $\begin{array}{r} 497 \\ -\ 113 \\ \hline \end{array}$ $\begin{array}{r} 95 \\ -\ 32 \\ \hline \end{array}$

2 $\begin{array}{r} 89 \\ -\ 70 \\ \hline \end{array}$ $\begin{array}{r} 72 \\ -\ 42 \\ \hline \end{array}$ $\begin{array}{r} 678 \\ -\ 372 \\ \hline \end{array}$ $\begin{array}{r} 87 \\ -\ 52 \\ \hline \end{array}$ $\begin{array}{r} 28 \\ -\ 17 \\ \hline \end{array}$

Rewrite each subtraction problem in column form. Be sure to put the correct number on top. Find each difference. Check your answers using addition.

3 256 − 134 = ______

4 59 − 32 = ______

5 87 − 45 = ______

6 242 − 101 = ______

If the numbers in a problem have a decimal point or a label, repeat it in your answer. If the top number has a dollar sign, repeat in the answer. Decimal points and labels line up vertically.

$$\begin{array}{r} \$2.56 \\ -\ 1.23 \\ \hline \$1.33 \end{array}$$

7 $\begin{array}{r} \$79 \\ -\ 30 \\ \hline \end{array}$

8 $\begin{array}{r} \$6.80 \\ -\ 3.20 \\ \hline \end{array}$

9 $\begin{array}{r} \$3.42 \\ -\ 2.40 \\ \hline \end{array}$

10 $\begin{array}{r} 78 \text{ feet} \\ -\ 32 \text{ feet} \\ \hline \end{array}$

Sometimes the difference in a column is zero. If the zero is at the left of a whole number, you do not have to write the zero.

$$\begin{array}{r} 67 \\ -\ 47 \\ \hline 20 \end{array} \qquad \begin{array}{r} 67 \\ -\ 62 \\ \hline 5 \end{array} \qquad \begin{array}{r} 67 \\ -\ 62 \\ \hline \text{not } 05 \end{array}$$

11 $\begin{array}{r} 49 \\ -\ 42 \\ \hline \end{array}$

12 $\begin{array}{r} 156 \\ -\ 106 \\ \hline \end{array}$

13 $\begin{array}{r} \$12.00 \\ -\ 10.00 \\ \hline \end{array}$

14 $\begin{array}{r} 236 \text{ feet} \\ -\ 235 \text{ feet} \\ \hline \end{array}$

The numbers **289** and **23** have a different number of digits. To subtract 23 from 289, line up the digits in the ones column. It is the same as you do in addition. You can think of an empty space in a column as a zero.

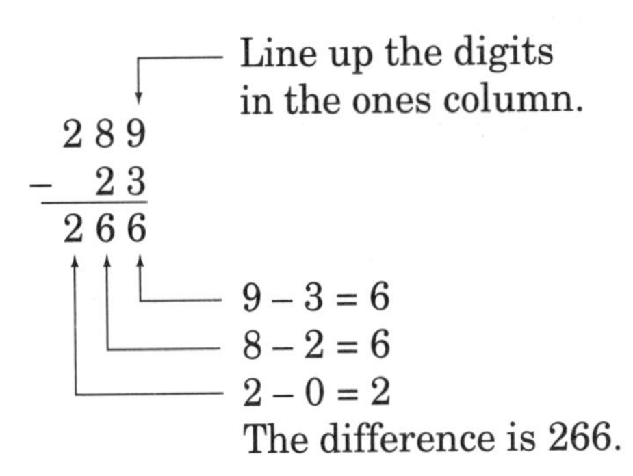

15 $\begin{array}{r} 67 \\ -\ 4 \\ \hline \end{array}$

16 $\begin{array}{r} 189 \\ -\ 75 \\ \hline \end{array}$

17 $\begin{array}{r} 79 \text{ in.} \\ -\ 6 \text{ in.} \\ \hline \end{array}$

18 $\begin{array}{r} 243 \\ -\ 21 \\ \hline \end{array}$

19 $\begin{array}{r} \$1.56 \\ -\ 0.30 \\ \hline \end{array}$

20 $\begin{array}{r} 590 \text{ meters} \\ -\ 40 \text{ meters} \\ \hline \end{array}$

21 $45 - 3 =$ ______

22 $135 - 25 =$ ______

23 $409 - 7 =$ ______

How To Borrow

In the ones column, 4 is larger than 3. When this happens, think of the top number 83 as **70 + 13** instead of **80 + 3.**

```
         ┌── 4 is larger than 3.
   83
 -  4
```

Thinking of 83 as 70 + 13 is called "borrowing" 10 from the 80.

```
   7 13
   8̸ 3̸  }← 83 = 70 + 13
 -   4
   7 9
   ↑ └── 13 − 4 = 9
   └──── 7 − 0 = 7
```

When you borrow, put a neat line through the number you are borrowing from. Write a new digit above it. (The new digit is **one less** than the old digit.) Move to the right and cross out the old digit. Then write a new number. (The new number is **ten more** than the old digit.)

PRACTICE

Solve each problem. Check your answers.

1 $\begin{array}{r} 32 \\ -\ 4 \\ \hline \end{array}$

2 $\begin{array}{r} 81 \\ -\ 5 \\ \hline \end{array}$

3 $\begin{array}{r} 73 \\ -\ 16 \\ \hline \end{array}$

4 $\begin{array}{r} 21 \\ -\ 19 \\ \hline \end{array}$

5 $\begin{array}{r} 56 \text{ votes} \\ -\ 27 \text{ votes} \\ \hline \end{array}$

6 $\begin{array}{r} 52 \text{ apples} \\ -\ 47 \text{ apples} \\ \hline \end{array}$

7 75 − 9 = ______

8 40 − 15 = ______

9 51 − 13 = ______

In this subtraction problem, 8 is larger than 5 in the tens column. Here, you borrow 10 tens (or 1 hundred) from 200. Write a new digit above the 2. Remember: the new digit is **one less** than the 2. Replace the digit to the right with a new number. The new number is **ten more** than the old digit.

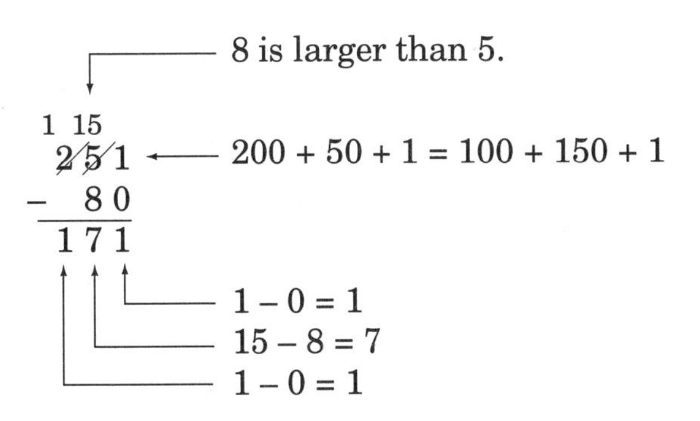

Solve each problem. Check your answers.

10 $\begin{array}{r} 452 \\ -\ 90 \\ \hline \end{array}$

11 $\begin{array}{r} 517 \text{ pounds} \\ -\ 32 \text{ pounds} \\ \hline \end{array}$

12 $\begin{array}{r} 206 \\ -\ 63 \\ \hline \end{array}$

13 $\begin{array}{r} 826 \text{ cups} \\ -\ 81 \text{ cups} \\ \hline \end{array}$

14 440 – 90 = ______

15 617 – 23 = ______

This problem needs two borrows. Start at the right and do one column at a time. *Remember: Once you have crossed out a number, you must use the new number written above it.*

12
4 2 11
5 3 1 ← 500 + 30 + 1 = 400 + 120 + 11
– 4 9
4 8 2

11 – 9 = 2
12 – 4 = 8
4 – 0 = 4

Solve each problem. Check your answers. *Being neat* is important, so you can tell which number goes with which column.

16 $\begin{array}{r} 532 \\ -\ 90 \\ \hline \end{array}$

17 $\begin{array}{r} 721 \\ -\ 382 \\ \hline \end{array}$

18 $\begin{array}{r} 457 \\ -\ 89 \\ \hline \end{array}$

19 $\begin{array}{r} 7{,}232 \\ -\ 75 \\ \hline \end{array}$

Borrowing from Zero

In this problem, you have to borrow from zero tens. To do that, you borrow 10 tens from the hundreds column. This example shows how.

Step 1: Take 1 hundred from the 4 hundreds.

```
 3 10
 4̸ 0̸ 7
-    9
```

Step 2: Take 1 ten from the 10 tens.

```
    9 17
 3 1̸0̸
 4̸ 0̸ 7̸
-     9
```

Step 3: Subtract.

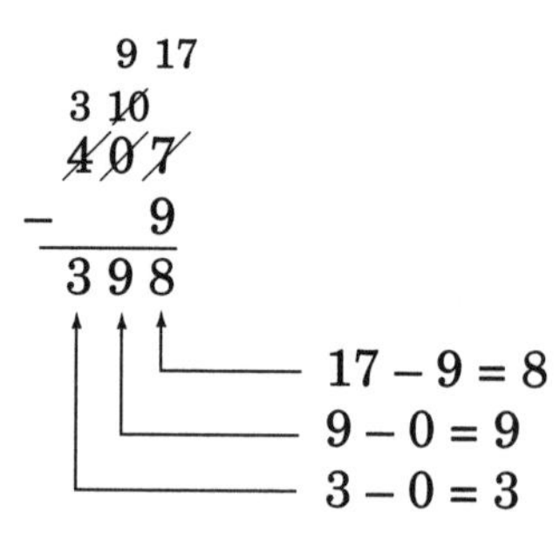

PRACTICE

Solve each problem. Use addition to check your answers.

1 503 − 9

2 803 − 15

3 700 − 16

4 $107 − 49

5 1,021 − 90

6 500 ounces − 27 ounces

7 1,040 − 903

8 3,050 − 721

9 100 − 9 = ____

10 400 − 35 = ____

11 507 − 19 = ____

12 1,005 − 209 = ____

Using Estimation To Check Subtraction

One way to estimate is to round each number. Then subtract using the rounded numbers.

You can use estimation with any type of problem: addition, subtraction, multiplication, or division.

Problem:

$$\begin{array}{r} 897 \\ -505 \\ \hline \end{array}$$

Round 897 to 900.
Round 505 to 500:

$$\begin{array}{r} 900 \\ -500 \\ \hline 400 \end{array}$$

The estimated difference is 400.

PRACTICE

Round each number to its left-most place. Then estimate the difference. ***You can review rounding on pages 13–14.***

1 $\begin{array}{r} 79 \\ -\ 52 \\ \hline \end{array}$

2 $\begin{array}{r} 83 \\ -\ 19 \\ \hline \end{array}$

3 $\begin{array}{r} 705 \\ -\ 689 \\ \hline \end{array}$

4 $\begin{array}{r} \$56 \\ -\ 41 \\ \hline \end{array}$

5 $\begin{array}{r} 121 \\ -\ 91 \\ \hline \end{array}$

6 $\begin{array}{r} 512 \text{ feet} \\ -\ 87 \text{ feet} \\ \hline \end{array}$

7 $\begin{array}{r} \$3.91 \\ -\ 2.03 \\ \hline \end{array}$

8 $\begin{array}{r} \$2.14 \\ -\ 1.21 \\ \hline \end{array}$

9 921 – 342 = ____

10 444 – 39 = ____

11 718 – 49 = ____

12 1,205 – 497 = ____

Checking Subtraction with First Digits

Here is another kind of estimation for subtraction. Write the first digit of each number, and replace the other digits with zeros. Then subtract. This is front-end estimation for subtraction.

Problem:

```
  981
- 216
```

Use the first digit of each number. Replace the other digits with zeros.

```
  900
- 200
  700
```

The estimate is 700.

You can tell whether the exact answer is more than or less than the estimate. Compare the numbers that you replaced with zeros. If the top number is larger than the bottom number, the exact answer is *greater than* your estimate. If the top number is less than the bottom number, the exact answer is **less than** your estimate.

Here is another example: 81 − 53. The estimated difference is 30.

Since the digit 1 from 81 is less than the digit 3 from 53, the exact answer is **less than** 30.

```
Estimate:  80
         - 50
           30
```

PRACTICE

Look at the first digit of each number. Use front-end estimation to estimate the answer to each problem. Then tell whether the exact answer is *less than* or *greater than* your estimate.

1 82 − 35

2 98 lb − 13 lb

3 679 − 114

4 710 − 683

5 305 − 179

Circle the letter for the *best* estimate to each problem.

6 95 − 31

A 60
B less than 60
C more than 60

7 $62 − $38 = ____

F less than $30
G less than $90
H more than $30

8 465 − 69 = ____

A 400
B less than 400
C more than 400

9 594 − 14

F exactly 500
G less than 500
H more than 500

Solving Word Problems

You can use the same five steps to solving most word problems. Review the five steps below. Then use them to solve the problems in the practice section below.

1. Identify the particular question in the problem.
2. Get all the information you need.
3. Set up the problem.
4. Solve it.
5. Check your work.

PRACTICE

What does each question ask for? Circle the letter for the best answer.
Remember: **If a problem asks you to take something away or to find a difference, it is a subtraction problem.**

1 Almost 130,000 people voted in a recent election. Saunders got 67,042 votes. Teller got 62,200 votes. How many more votes did Saunders get than Teller?

A an exact sum
B an exact difference
C an estimated difference
D an estimated sum

2 A 20-inch TV is on sale for about $290. That is $60 less than the original price. About how much does the TV usually cost?

F an exact sum
G an exact difference
H an estimated difference
J an estimated sum

3 At the end of Luca's diet, he weighed 183 pounds. In the last six months, he has gained back 17 pounds. How much does Luca weigh now?

A an exact sum
B an exact difference
C an estimated difference
D an estimated sum

4 Gina is saving up to buy a couch. It costs just over $830. So far, she has saved almost $550. About how much more does she need to save?

F an exact sum
G an exact difference
H an estimated difference
J an estimated sum

In each problem, circle the numbers you need to solve that problem. Then set up and solve the problem. *Remember:* Add and subtract *like things* only.

Sample Almost 130,000 people voted in a recent election. Saunders got 67,042 votes. Teller got 62,200 votes. How many more votes did Saunders get than Teller?

$$\begin{array}{r} 67{,}042 \\ -\ 62{,}200 \\ \hline \end{array}$$

5 A 20-inch TV is on sale for about $290. That is $60 less than the original price. About how much does the TV usually cost?

6 At the end of Luca's diet, he weighed 183 pounds. In the last six months, he has gained back 17 pounds. How much does Luca weigh now?

7 Gina is saving up to buy a couch. It costs just over $830. So far, she has saved almost $550. About how much more does she need to save?

8 Juana has a coupon for $0.20 off the price of a 16-ounce box of Wheatos. The box is marked $4.10. How much will Juana pay?

Information is missing in these word problems. Use the blank lines to tell what information is needed to solve the problem.

9 Dominique is driving 320 miles to see her mother. So far she has spent 3 hours on the road. How many more miles does she have to go?

You also need to know

____

____

10 Enrique took his family to the circus. He took $30.00. Tickets were $5.00 apiece. How much money does he have left to spend on treats?

You also need to know

____

____

11 Eighty-nine people work for Ron's company. Twenty-five of them have technical degrees. How many of them did not finish high school?

You also need to know

____

____

12 Mr. Swenson makes $610 a week. How much more does he make than his younger brother?

You also need to know

____

____

Follow the five steps on page 41 to solve these problems. Be sure to show all your work and how you checked your answers.

13 Emilio is 42 years old. His son just turned 17. How old was Emilio when his son was born?

14 Rochelle and her roommate live in an apartment. The rent is $625 per month. Rochelle pays $310 per month. How much does her roommate pay?

Subtraction Skills Practice

Circle the letter for the best answer to each problem.
Try crossing out any unreasonable answers before you start to work.

1

$$\begin{array}{r} 65 \\ -\ 42 \\ \hline \end{array}$$

A 43
B 42
C 22
D 23
E None of these

2

$$\begin{array}{r} 42 \\ -\ 33 \\ \hline \end{array}$$

F 9
G 11
H 19
J 75
K None of these

3

$$\begin{array}{r} 97 \\ -\ 63 \\ \hline \end{array}$$

A 32
B 34
C 35
D 45
E None of these

4

$$\begin{array}{r} 160 \\ -\ 80 \\ \hline \end{array}$$

F 60
G 40
H 80
J 180
K None of these

5

922 – 13 = ____

A 919
B 909
C 809
D 819
E None of these

6

50 – 30 = ____

F 10
G 20
H 30
J 47
K None of these

7

$$\begin{array}{r} 932 \\ -\ 131 \\ \hline \end{array}$$

A 819
B 929
C 921
D 829
E None of these

Use this information to do Numbers 8 through 10.

DeShawn is going on a 150-mile car trip. He starts with $25.00 to spend on gas and food.

8 After DeShawn drives 70 miles, he stops to eat. How can you find out how much farther he has to go?

F Add.
G Divide.
H Multiply.
J Subtract.

9 DeShawn spends $14.50 on dinner. How much money does he have left for gas?

A $10.50
B $21.50
C $11.50
D $39.50

10 When the trip is over, DeShawn has $6.25 left. How much money did he spend?

F $18.25
G $19.25
H $18.75
J $19.75

11

91 − 4 = ____

A 86
B 88
C 85
D 84
E None of these

12

7 − 3 = ____

F 3
G 5
H 4
J 2
K None of these

13

```
  38
− 2
```

A 36
B 40
C 35
D 37
E None of these

14

85 − 25 = ____

F 6
G 60
H 65
J 50
K None of these

15

```
  300
− 67
```

A 232
B 233
C 333
D 243
E None of these

16

$5.00 − $3.50 = ____

F $1.50
G $0.50
H $2.50
J $2.00
K None of these

17

36 − 17 = ____

A 21
B 19
C 29
D 31
E None of these

18 Which of these, if any, has the same value as 57 − 12 ?

F 57 − 12 − 0
G 12 − 57
H 12 + 57
J 57 − 12 − 1
K None of these

19 Which of these, if any, is false?

A 12 − 0 = 12
B 20 − 5 − 3 = 20 − 8
C 156 − 156 = 0
D If 872 − 529 = 343, then 343 + 529 = 872.
E All of the statements are true.

20 What number goes in the box to make the second number sentence true?

158 + 32 = 190

190 − ☐ = 158

F 32
G 158
H 190
J 42
K None of these

Multiplication

Basic Concepts

For some problems, you add the same number repeatedly. A shortcut for repeated addition is **multiplication.**

Wayne bought tickets for 4 people. The tickets cost 7 dollars each. How much money did he spend?

7 dollars	**7 dollars**
7 dollars	**× 4**
7 dollars	**28 dollars**
+ 7 dollars	
28 dollars	*Think:* 4 sevens is 28.

- Some of the terms that signal "multiplication" are *times,* and *multiplied by,* and *product.* Here are some examples:

  What is 7 *times* 3?
  How much is 5 *multiplied by* 5?
  What is the *product* of 2 and 10?

- A multiplication problem can be written from left to right in a row, or from top to bottom in columns. The symbol "×" tells you to multiply. In a multiplication problem, the result is called a **product.**

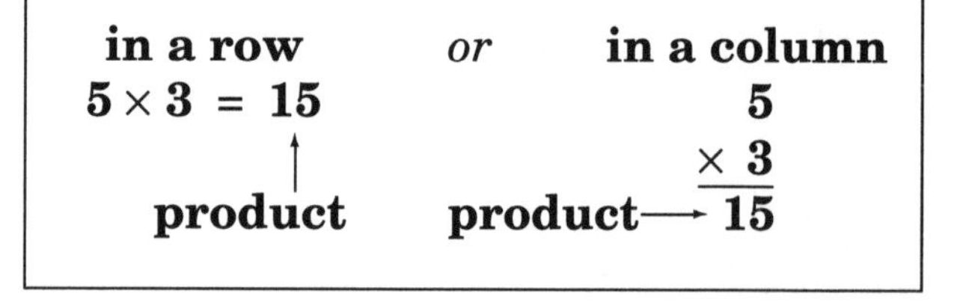

- You can change the order of the numbers when you multiply. The order of the numbers does not change the product.

  The value of **3** × **4** is the same as the value of **4** × **3.**

- If you multiply a number by 1, you do not change its value. *Think:* one times 23 is 23.

  $1 \times 23 = 23$
  $23 \times 1 = 23$

- If you multiply any number times zero, the product is zero. *Think:* zero times 71 is zero.

  $71 \times 0 = 0$
  $0 \times 71 = 0$

- To multiply several numbers together, multiply two numbers at a time. You can pick the numbers in any order.

$$2 \times 3 \times 3 = (2 \times 3) \times 3 = (6) \times 3 = 18$$

or

$$2 \times 3 \times 3 = 2 \times (3 \times 3) = 2 \times (9) = 18$$

PRACTICE

Fill in each blank.

1 $262 \times 1 =$ ______

2 $697 \times 0 =$ ______

3 $15 + 15 + 15 + 15 =$ ______

(Write this as a multiplication problem.)

4 $61 \times 3 =$ ______

(Write this as an addition problem.)

5 $2 \times 2 \times 3 =$ ______ $\times$ ______
$=$ ______ $\times$ ______ } Write this two different ways.

Write T for *true* or F for *false*.

6 52×13 has the same value as 13×52. ______

7 $17 \times 4 = 4 + 4 + 4 + 4$ ______

8 $3 \times 12 = 12 + 12 + 12$ ______

9 $12 \times 0 = 12$ ______

10 $43 \times 1 = 1$ ______

11 $142 \times 142 = 0$ ______

12 $3 \times 2 \times 2$ has the same value as 6×2. ______

13 $7 \times 2 \times 4$ has the same value as 7×8. ______

Basic Multiplication Facts

The multiplication problems 1×1 through 9×9 are the **basic multiplication facts.** You should have all these facts in your memory.

PRACTICE

Try the problems on this page. If you make more than a few mistakes, erase your answers. Then do the problems again.

1 $2 \times 6 =$ ______ $4 \times 4 =$ ______ $2 \times 4 =$ ______

2 $2 \times 3 =$ ______ $8 \times 4 =$ ______ $6 \times 6 =$ ______

3 $6 \times 5 =$ ______ $5 \times 3 =$ ______ $7 \times 6 =$ ______

4 $5 \times 2 =$ ______ $7 \times 4 =$ ______ $8 \times 9 =$ ______

5 $\begin{array}{r} 4 \\ \times 3 \\ \hline \end{array}$ $\begin{array}{r} 2 \\ \times 2 \\ \hline \end{array}$ $\begin{array}{r} 7 \\ \times 3 \\ \hline \end{array}$ $\begin{array}{r} 3 \\ \times 9 \\ \hline \end{array}$

6 $\begin{array}{r} 7 \\ \times 7 \\ \hline \end{array}$ $\begin{array}{r} 8 \\ \times 3 \\ \hline \end{array}$ $\begin{array}{r} 3 \\ \times 3 \\ \hline \end{array}$ $\begin{array}{r} 4 \\ \times 5 \\ \hline \end{array}$

7 $\begin{array}{r} 5 \\ \times 9 \\ \hline \end{array}$ $\begin{array}{r} 7 \\ \times 5 \\ \hline \end{array}$ $\begin{array}{r} 2 \\ \times 7 \\ \hline \end{array}$ $\begin{array}{r} 4 \\ \times 6 \\ \hline \end{array}$

8 $\begin{array}{r} 8 \\ \times 2 \\ \hline \end{array}$ $\begin{array}{r} 2 \\ \times 9 \\ \hline \end{array}$ $\begin{array}{r} 8 \\ \times 5 \\ \hline \end{array}$ $\begin{array}{r} 8 \\ \times 8 \\ \hline \end{array}$

9 $\begin{array}{r} 9 \\ \times 4 \\ \hline \end{array}$ $\begin{array}{r} 7 \\ \times 8 \\ \hline \end{array}$ $\begin{array}{r} 5 \\ \times 5 \\ \hline \end{array}$ $\begin{array}{r} 6 \\ \times 8 \\ \hline \end{array}$

10 $\begin{array}{r} 7 \\ \times 9 \\ \hline \end{array}$ $\begin{array}{r} 3 \\ \times 6 \\ \hline \end{array}$ $\begin{array}{r} 9 \\ \times 9 \\ \hline \end{array}$ $\begin{array}{r} 6 \\ \times 9 \\ \hline \end{array}$

Multiplying by a One-Digit Number

Here is a multiplication problem in column form. The larger number is on top and the smaller number on the bottom. For multiplication, the digits line up at the right.

Multiply the bottom number by each digit in the top number. Work from right to left.

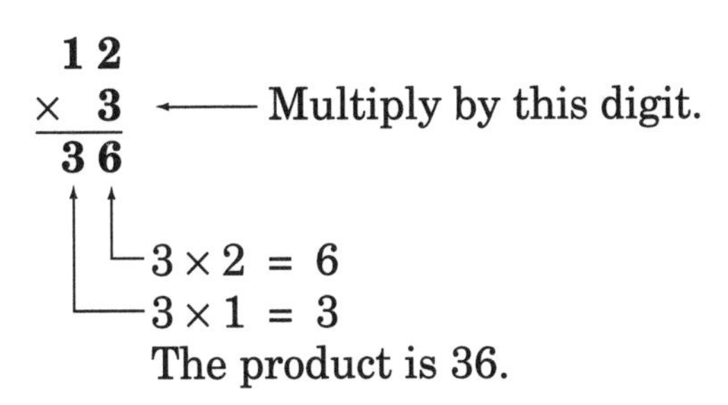

PRACTICE

Find each product. Be sure to write every digit, even if it is a zero.

1 $\begin{array}{r} 23 \\ \times 3 \\ \hline \end{array}$ $\begin{array}{r} 141 \\ \times 2 \\ \hline \end{array}$ $\begin{array}{r} 121 \\ \times 2 \\ \hline \end{array}$ $\begin{array}{r} 42 \\ \times 4 \\ \hline \end{array}$ $\begin{array}{r} 31 \\ \times 7 \\ \hline \end{array}$

2 $\begin{array}{r} 103 \\ \times 3 \\ \hline \end{array}$ $\begin{array}{r} 53 \\ \times 3 \\ \hline \end{array}$ $\begin{array}{r} 10 \\ \times 7 \\ \hline \end{array}$ $\begin{array}{r} 64 \\ \times 2 \\ \hline \end{array}$ $\begin{array}{r} 82 \\ \times 4 \\ \hline \end{array}$

Rewrite each multiplication problem in column form. Be sure the digits line up at the right. Find each product.

3 $134 \times 2 =$ ______

4 $2 \times 54 =$ ______

5 $4 \times 80 =$ ______

6 $210 \times 3 =$ ______

When a Column Product Is More Than 9

In the multiplication problem below, the product in the right column is a 2-digit number. You write the digit 8, from the ones place, as part of your answer. Then you write the digit 2, from the tens place, above the next digit. The digit 2 is called the *carry number.* Finally, multiply 7 times 5, then add the carry number.

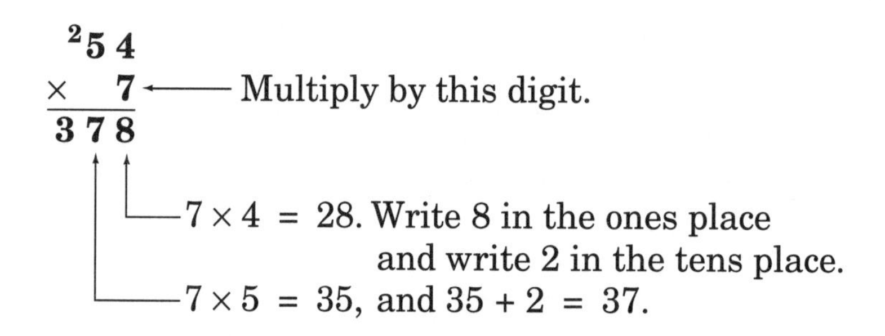

PRACTICE

Solve each problem. Remember, multiply before you add the "carry" digit. Be sure the problem is in column form before you multiply.

1 15×4

2 27×3

3 23×6

4 12×9

5 56×2

6 25×2

7 $75 \times 3 =$ ______

8 $14 \times 6 =$ ______

9 $35 \times 3 =$ ______

If one of the numbers in a problem has a label, repeat that label in your answer.

10 23 pounds × 9

11 27 dollars × 3

12 45 inches × 2

13 37 feet × 3

Here is another set of problems. Remember these key ideas:

1 Whenever a number is multiplied by 0, the product is 0.
2 Always multiply *before* you add a carry digit.

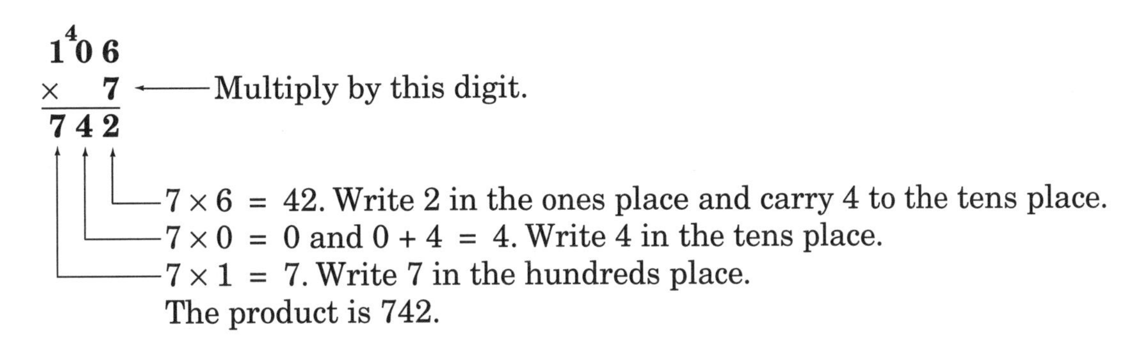

Be sure each problem is in column form before you multiply.

14 $\begin{array}{r} 105 \\ \times\ 9 \\ \hline \end{array}$

15 $\begin{array}{r} \$207 \\ \times\ 3 \\ \hline \end{array}$

16 $\begin{array}{r} 400 \\ \times\ 3 \\ \hline \end{array}$

17 $\begin{array}{r} 561 \\ \times\ 3 \\ \hline \end{array}$

18 705 × 2 = ____

19 700 × 4 = ____

Each of these problems has more than one carry number. Work in column form, from right to left, one column at a time. Write neatly.

20 $\begin{array}{r} 125 \\ \times\ 5 \\ \hline \end{array}$

21 $\begin{array}{r} 137 \\ \times\ 3 \\ \hline \end{array}$

22 $\begin{array}{r} \$147 \\ \times\ 3 \\ \hline \end{array}$

23 $\begin{array}{r} 156 \\ \times\ 3 \\ \hline \end{array}$

24 $\begin{array}{r} 255 \\ \times\ 2 \\ \hline \end{array}$

25 $\begin{array}{r} 237 \\ \times\ 4 \\ \hline \end{array}$ feet

26 650 × 2 = ____

27 420 × 6 = ____

28 189 × 2 = ____

Using Estimation When You Multiply

To estimate this product, round the top number to the nearest hundred. Then multiply.

Be sure to write the correct number of zeros in your estimate. If your estimate does not have enough digits, it will not be useful or correct.

Problem: 734×7

Round 734 to 700. Then multiply.

$$\begin{array}{r} 700 \\ \times \quad 7 \\ \hline 4{,}900 \end{array}$$

An estimate is 4,900.

PRACTICE

Estimate each product by rounding the top number to the nearest ten, hundred, or thousand. ***Hint:*** **Your estimate should have at least as many zeros as the rounded number.**

1 79×2

2 93×9

3 27×9

4 98×3 ($98)

5 212×6

6 575×8

7 423×5 ($423)

8 $6{,}598 \times 2$

9 $921 \times 3 =$ ______

10 $314 \times 9 =$ ______

11 $1{,}918 \times 6 =$ ______

12 $1{,}092 \times 7 =$ ______

Solving Word Problems

An important part of solving any word problem is deciding whether to add, subtract, multiply, or divide. Look for these clues.

- If you **put** amounts **together,** you add.
- If you **compare** one amount to another, you subtract or divide.
- If you **take something away** from something else, you subtract.
- If you deal with **some number of groups,** you multiply or divide.

Signal Words

Addition	Subtraction	Multiplication	Division
plus sum total added to altogether in all combined increased by	minus difference take away subtract left over how much change how much more than how much less than decreased by	times multiplied by product twice, three times two or more of something apiece each	This is in the next section.

PRACTICE

For each problem, tell whether you will add, subtract, or multiply to find the answer. You *do not* have to find the actual answer.

1 In May, Jackie's rent is increasing \$30. She pays \$515 now. How much will her new rent be?

A Add.
B Subtract.
C Multiply.

2 Carmella weighs 135 pounds. When she is holding her cat, her scale shows 146 pounds. How much does Carmella's cat weigh?

F Add.
G Subtract.
H Multiply.

3 Last week, Lionel sold five times as many shoes as Wayne. Wayne sold 27 pairs of shoes. How many pairs of shoes did Lionel sell?

A Add.
B Subtract.
C Multiply.

4 Hector bought 6 shirts for \$25 each. They were originally marked \$30 each. How much did he pay?

F Add.
G Subtract.
H Multiply.

Use this information for problems 5 and 6.

Premium unleaded gasoline at the Aramco station is $1.59 a gallon. Super unleaded gasoline is $1.43 a gallon. Regular unleaded gasoline is $1.29 a gallon.

5 How much would 9 gallons of premium gasoline cost?

A Add.
B Subtract.
C Multiply.

6 How much more is a gallon of premium unleaded gasoline than a gallon of regular unleaded gasoline?

F Add.
G Subtract.
H Multiply.

Set up each problem. You do not have to solve it. Remember: In a subtraction problem, the number you are *subtracting from* goes on top. The number you are *taking away* goes on the bottom.

7 In May, Jackie's rent is increasing $30. She pays $515 now. How much will her new rent be?

8 Carmella weighs 135 pounds. When she is holding her cat, her scale shows 146 pounds. How much does Carmella's cat weigh?

9 Last week, Lionel sold five times as many shoes as Wayne. Wayne sold 27 pairs of shoes last week. How many pairs of shoes did Lionel sell?

10 Hector bought 6 shirts for $25 each. They were originally marked $30 each. How much did he pay?

Use this information for problems 11 and 12.

Premium unleaded gasoline at the Aramco station is $1.59 a gallon. Super unleaded gasoline is $1.43 a gallon. Regular unleaded gasoline is $1.29 a gallon.

11 How much would 9 gallons of premium unleaded gasoline cost?

12 How much more is a gallon of premium unleaded gasoline than a gallon of regular unleaded gasoline?

Information is missing in these word problems. On each blank line, tell what information is needed to solve the problem.

13 Cleo bought 3 boxes of apples. About how many apples did she buy?

You also need to know

____

____

14 Randy has been hiking at a steady pace for three days. About how far has he hiked?

You also need to know

____

____

15 Exactly 27 students are in each class at Woodward Elementary School. About how many students attend the school?

You also need to know

____

____

Solve these problems on your own. Be sure to show all your work and show how you checked your answers.

16 Elaine bought 32 raffle tickets. They cost 2 dollars each. How much did she spend?

17 The Statue of Liberty is 152 feet tall. The Washington Monument is 403 feet taller than the Statue of Liberty. How tall is the Washington Monument?

18 A cheetah can run 70 miles an hour. A man can run 28 miles an hour. How much faster is the cheetah than the man?

19 Chung Li can build 4 spice racks in a day. How many spice racks can he build in 12 days?

Multiplication Skills Practice

Circle the letter for the best answer to each problem.

1

$$\begin{array}{r} 22 \\ \times\ 3 \\ \hline \end{array}$$

A 55
B 66
C 25
D 62
E None of these

2

$$\begin{array}{r} 62 \\ \times 3 \\ \hline \end{array}$$

F 186
G 185
H 126
J 96
K None of these

3

$310 \times 4 =$ ____

A 124
B 314
C 1,240
D 3,104
E None of these

4

$$\begin{array}{r} 24 \\ \times 3 \\ \hline \end{array}$$

F 27
G 72
H 62
J 612
K None of these

5

$$\begin{array}{r} 35 \\ \times 7 \\ \hline \end{array}$$

A 42
B 245
C 215
D 222
E None of these

6

$$\begin{array}{r} 24 \\ \times 9 \\ \hline \end{array}$$

F 193
G 186
H 216
J 33
K None of these

Kavita sells wedding cakes. To make them, she triples this recipe. Study the recipe. Then do Numbers 7 through 9.

4 cups flour
1 cup sugar
1 cup butter
3 eggs
1 tablespoon baking powder
juice from one lemon

Sift the flour and sugar. Mix in the eggs, baking powder, butter, and juice. Bake for 45 minutes at 350 degrees.

7 How much flour does Kavita need to make one wedding cake?

A 4 cups
B 8 cups
C 10 cups
D 12 cups

8 It takes Kavita thirty minutes to mix a cake, forty-five minutes to bake it, and two hours to decorate it. How can she find out how much time it takes in all?

F Add.
G Divide.
H Multiply.
J Subtract.

9 Kavita sells an average of 9 cakes a month. At that rate, how many will she sell in a year?

A 90
B 98
C 108
D 210

10

$$\begin{array}{r} 6 \\ \times\ 5 \\ \hline \end{array}$$

F 25
G 30
H 35
J 36
K None of these

11

$$\begin{array}{r} 420 \\ \times\ \ 3 \\ \hline \end{array}$$

A 1,260
B 126
C 180
D 1,250
E None of these

12

$22 \times 0 =$ ____

F 0
G 1
H 22
J 220
K None of these

13

$$\begin{array}{r} 13 \\ \times\ 5 \\ \hline \end{array}$$

A 55
B 58
C 515
D 65
E None of these

14

$104 \times 4 =$ ____

F 56
G 4,016
H 416
J 560
K None of these

15

$6 \times 3 \times 1 =$ ____

A 18
B 19
C 9
D 24
E None of these

16

$6 \times 6 =$ ____

F 66
G 12
H 30
J 36
K None of these

17 Which of these, if any, has the same value as 12×5?

A one-fifth of twelve
B $12 - 5$
C 5×12
D $12 \times 5 \times 0$
E None of these

18 Which of these, if any, has the same value as 30×2?

F $30 \times 1 \times 1$
G $30 + 30$
H $30 + 2$
J 32
K None of these

19 Which of these, if any, equals 5?

A 5×0
B 0×5
C 5×1
D 5×5
E None of these

Division

Basic Concepts

For some problems, you separate a group of objects into smaller groups. This is called **division.**

Wayne bought tickets for 4 people. He spent \$32.00. How much did each ticket cost?	Traci has 12 dollars. Hot dogs are 2 dollars each. How many hot dogs can she buy?

- Two symbols for division are ÷ and $\overline{)}$. The answer in a division problem is called a **quotient.**

- Here are some problems using the division sign ÷.
  Twelve divided by 4 is "12 ÷ 4."
  That means "Divide 12 by 4."
  12 ÷ 4 = 3, so the quotient is 3.

  3 divided into 18 is "$3\overline{)18}$."
  That means "Divide 18 by 3."
  $\begin{array}{r} 6 \\ 3\overline{)18} \end{array}$ so the quotient is 6.

- The order of the numbers in a division problem is important.

  12 ÷ 4 is the same as **$4\overline{)12}$.**

  12 ÷ 4 is *not* the same as **4 ÷ 12.**

- If you divide a number by 1, you do not change the value. *Think:* If you divide 92 into 1 group, that group has to contain 92 objects.

  92 ÷ 1 = 92 $\qquad \begin{array}{r} 92 \\ 1\overline{)92} \end{array}$

- If you divide any number by itself, the quotient is 1. *Think:* If you divide 72 objects into 72 groups, each group would have 1 object.

  71 ÷ 71 = 1 $\qquad \begin{array}{r} 1 \\ 71\overline{)71} \end{array}$

- If you divide zero by any (nonzero) number, the quotient is zero. You can think about dividing 0 objects into 15 groups. Each group has 0 objects. You *cannot* divide any number by zero.

  0 ÷ 71 = 0

  71 ÷ 0 does not have any meaning in math.

- Division and multiplication are the opposites of each other.

  Multiply by 3.
  12 × 3 = 36

  Divide by 3.
  36 ÷ 3 = 12

PRACTICE

1 **Circle the letter for each word problem that calls for division. *Do not* try to solve the problems.**

A Chris bought 8 pounds of tomatoes for $4.00. How much did he pay per pound?

B Chris paid for the tomatoes with a $5.00 bill. How much change should he get back?

C Maria is buying gifts. She spent $23 on her mother, $27 on her sister, and $45 on her son. How much did she spend in all?

D Garth and his sister made $630 painting a house. They will divide the money evenly in half. How much money will each person get?

Fill in each blank.

2 $36 \div 1 =$ ______

3 $0 \div 105 =$ ______

4 If $42 \times 2 = 84$, then $84 \div 2 =$ ______.

5 If $75 \div 3 = 25$, then $25 \times 3 =$ ______.

6 $142 \div 142 =$ ______

7 You can write $16 \div 4$ with a division bracket as ______

8 If $3 \times 31 = 93$, then

____

(Write a division problem.)

9 If $224 \div 4 = 56$, then

____

(Write a multiplication problem.)

Write T for *true* or F for *false*.

10 $12 \div 5$ has the same value as $5 \div 12$. ______

11 $63 \div 3$ has the same value as $63\overline{)3}$. ______

12 $43 \div 0 = 0$ ______

13 $142 \div 142 = 0$ ______

14 If $144 \div 12 = 12$, then $12 + 12 = 124$. ______

Basic Division Facts

Since $4 \times 5 = 20$, then
$20 \div 5 = 4$ and
$20 \div 4 = 5$.

Each basic division fact is related to a multiplication fact. You should be able to answer any basic fact problem quickly and accurately.

PRACTICE

Do the problems below. If you make more than a few mistakes, try them again.

1	$12 \div 6 =$ ______	$16 \div 4 =$ ______	$8 \div 4 =$ ______	
2	$6 \div 3 =$ ______	$32 \div 4 =$ ______	$36 \div 6 =$ ______	
3	$30 \div 5 =$ ______	$15 \div 3 =$ ______	$9 \div 3 =$ ______	
4	$10 \div 2 =$ ______	$28 \div 4 =$ ______	$0 \div 5 =$ ______	
5	$12 \div 3 =$ ______	$4 \div 2 =$ ______	$21 \div 3 =$ ______	$27 \div 3 =$ ______
6	$49 \div 7 =$ ______	$24 \div 3 =$ ______	$42 \div 6 =$ ______	$20 \div 5 =$ ______
7	$45 \div 5 =$ ______	$35 \div 5 =$ ______	$14 \div 7 =$ ______	$24 \div 6 =$ ______
8	$16 \div 2 =$ ______	$18 \div 2 =$ ______	$40 \div 5 =$ ______	$64 \div 8 =$ ______
9	$36 \div 4 =$ ______	$56 \div 8 =$ ______	$25 \div 5 =$ ______	$48 \div 8 =$ ______
10	$63 \div 9 =$ ______	$18 \div 3 =$ ______	$81 \div 9 =$ ______	$54 \div 9 =$ ______

Using the Division Symbol $\overline{)}$

To set up a division problem using the symbol $\overline{)}$, the number you are **dividing** goes *inside* the bracket. This number is called the **dividend.** The number you are **dividing by** goes to the *left* of the bracket. It is called the **divisor.**

Write $69 \div 3$ as $3\overline{)69}$

↑ **divisor** ↓ ↑ **dividend** ↓

Write $56 \div 7$ as $7\overline{)56}$

PRACTICE

Rewrite each problem using the division symbol $\overline{)}$. You *do not* have to solve the problem.

1 $64 \div 2 =$ ______

2 $88 \div 8 =$ ______

3 $93 \div 3 =$ ______

4 What is 55 divided by 5?

5 What is sixty-three divided by 3?

6 What is 28 inches divided by 4?

7 $48 \div 4 =$ ______

8 $28 \div 2 =$ ______

9 $39 \div 3 =$ ______

10 Amelia must divide 84 dollars among 4 people. How much money will each person get?

11 Frank paid $46 for 2 picture frames. How much did he pay for each frame?

Dividing by a One-Digit Number

In this problem, the number you are **dividing by** (the **divisor**) has one digit. Start at the left of the dividend 96. First, divide 3 into 9. then divide 3 into 6.

Divide 3 into 9: $9 \div 3 = 3$.
Divide 3 into 6: $6 \div 3 = 2$.

$3\overline{)96}$ with quotient 32. The quotient is 32.

To check your work, multiply your answer (the quotient) by the divisor.

$$\begin{array}{r} 32 \\ \times\ 3 \\ \hline 96 \end{array}$$

32 ← quotient
3 ← divisor

PRACTICE

Find each quotient. Use multiplication to check your work. *Remember:* If you divide zero by any (nonzero) number, the result is zero.

1 $3\overline{)93}$

2 $5\overline{)55}$

3 $2\overline{)84}$

4 $3\overline{)60}$

5 $4\overline{)484}$

6 $2\overline{)224}$

7 $2\overline{)28}$

8 $4\overline{)\$80}$

9 $3\overline{)69}$

Rewrite each problem using the division symbol $\overline{)\ }$. Solve each division problem. Be sure to check your answers. *Remember: The* number you are dividing *into* goes inside the bracket.

10 Find 64 divided by 2.

11 $105 \div 5 =$ ____

12 Cecilia has $250 to spend on presents. She wants to spend the same amount for each of 5 people. How much can she spend on each person?

When a Dividend Digit Is Too Small

In this problem, the digit 2 inside the division symbol is smaller than the digit 3 outside the symbol.

To start, look at the number formed by the first two digits inside the division symbol. That number is 21, so divide 3 into 21.

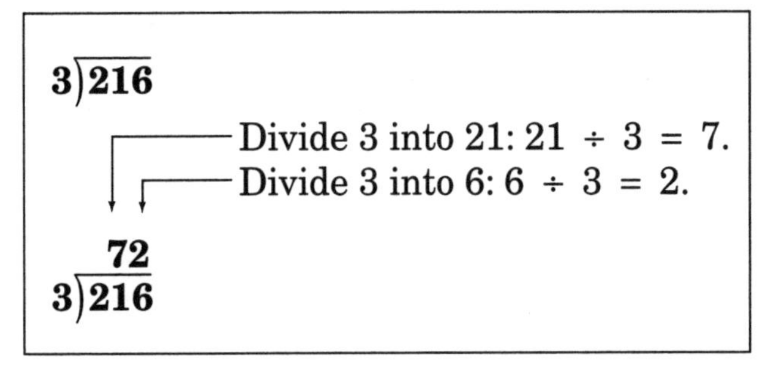

PRACTICE

Solve each problem. Check your answers.

1 $3\overline{)159}$

2 $5\overline{)405}$

3 $8\overline{)168}$

4 150 divided by 3

5 $6\overline{)306}$

6 $4\overline{)248}$

7 $148 \div 2 =$ ______

8 $5\overline{)100}$

In this problem, the second digit in 812 is too small. For this, write a zero in the quotient. Then divide 4 into the two-digit number 12.

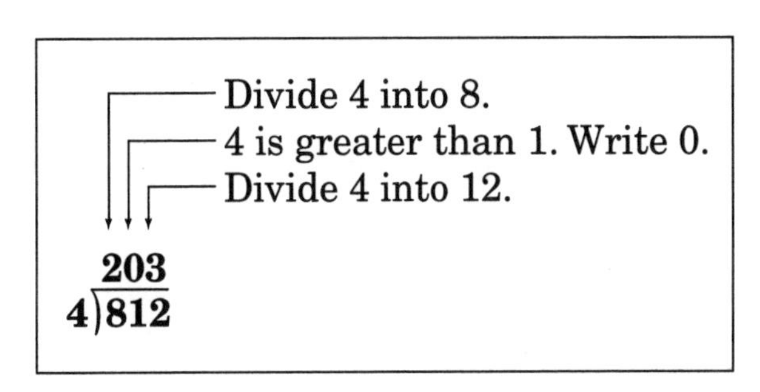

9 $3\overline{)315}$

10 $5\overline{)540}$

11 $8\overline{)448}$

12 615 divided by 3

13 $6\overline{)630}$

14 $4\overline{)416}$

15 $610 \div 2 =$ ______

16 $5\overline{)510}$

Dividing with a Remainder

For the division problem $2\overline{)9}$, the answer is not a whole number. If you divide 9 objects into 2 groups, each group has 4 objects, with 1 object left over. That left over number is called a **remainder.**

9 ÷ 2 = 4 r 1

For the division problem *9 divided by 2,* the quotient is *4 with a* ***remainder*** *of 1.*
In the problem below, the number 3 does not divide evenly into the number 17.
The quotient is "5 remainder 2."

```
     ┌──17 ÷ 3 is a little more than 5. Write 5.
     ↓
     5
  3)17
    15  ←Multiply: 5 × 3 = 15.
     2  ←Subtract: 17 − 15 = 2.
         The quotient is 5 r 2.
```

The number 5 is the *largest possible whole number* for "17 divided by 3." The number 2 is the difference when 15 is subtracted from 17.

PRACTICE

Solve each problem. *Hint:* Any remainder is always *smaller* than the divisor outside the division symbol.

1 $3\overline{)7}$

2 $5\overline{)16}$

3 $7\overline{)24}$

4 $6\overline{)34}$

5 $8\overline{)29}$

6 $10\overline{)78}$

7 22 ÷ 3 = ______

8 51 ÷ 6 = ______

9 38 ÷ 9 = ______

10 39 divided by 6

11 If you divide 51 coins among 7 people, how many coins does each person get? How many are left over?

Check each problem on this page. Here is an example of the three steps in a check.

1 Multiply the quotient 5 and the divisor 3. — 5 × 3 = 15
2 Add the remainder 2. — 15 + 2 = 17
3 The result should be the same as the dividend 17.

```
   5 r 2
3)1 7
  1 5
    2
```

Writing the Steps of a Division Problem

In this problem, the divisor 2 does not evenly divide the first digit in the dividend. The largest possible whole number is 4. Multiply 4 and 2, and subtract the result from 9. Then "bring down" 6, the next digit inside the division symbol.

$$\begin{array}{r} 48 \\ 2\overline{)96} \\ \underline{8} \\ 16 \\ \underline{16} \\ 0 \end{array}$$

Divide 2 into 16. Write the 8 in the quotient. Multiply 8 and 2, and subtract the result from 16. The difference is zero, so the problem has no remainder.

This example shows a step-by-step method for solving such problems.

$$\begin{array}{r} 48 \\ 2\overline{)96} \\ \underline{8} \\ 16 \\ \underline{16} \\ 0 \end{array}$$

1 Divide 9 by 2. Write the 4 above the 9.
2 Multiply: 4 and 2. Write the 8 below the 9 and subtract.
3 *Bring down* the next digit 6.
4 Divide 16 by 2. Write the 8.
5 Multiply 8 and 2.
6 Subtract: 16 − 16 = 0, so the problem has no remainder.

PRACTICE

Solve each problem. Be sure it is written using the division bracket $\overline{)}$. Check your work. It is important to write every step, including the subtractions.

1 $2\overline{)36}$

2 $4\overline{)56}$

3 $4\overline{)60}$

4 $3\overline{)84}$

5 $5\overline{)90}$

6 $8\overline{)96}$

7 $2\overline{)138}$

8 $6\overline{)726}$

9 222 ÷ 3 = ______

10 165 ÷ 5 = ______

11 58 ÷ 2 = ______

12 231 divided by 7

13 52 divided by 4

14 A 137-inch board must be cut into 9-inch pieces. How many 9-inch pieces will there be? How much will be left over?

Mixed Practice

Solve each division problem on this page. Check your work. You may need to review some of the earlier sections on division.

1 $4\overline{)844}$

2 $81 \div 9 =$ ____

3 $2\overline{)602}$

4 $72 \div 8 =$ ____

5 $5\overline{)180}$

6 $210 \div 7 =$ ____

7 $3200 \div 8 =$ ____

8 $5\overline{)510}$

9 $5\overline{)45}$

10 $2\overline{)13}$

11 $5\overline{)805}$

12 $915 \div 3 =$ ____

13 $150 \div 5 =$ ____

14 What is 64 divided by 4?

15 What is 132 divided by 2?

16 What is 608 divided by 6?

17 Colleen has 25 dollars. At a flea market, picture frames cost 5 dollars each. How many picture frames can she buy?

18 Timothy painted 40 feet of fencing in 8 hours. About how many feet of fencing did he paint per hour?

19 There are 8 children in Latisha's daycare. Last month, she spent $328 on supplies. About how much did she spend on supplies per child?

20 Winny has $35. Candy is $2 per bag. How many bags of candy can she buy?

Using Estimation with Division

To estimate an answer to a division problem, find the first digit in the quotient. Look at the rest of the digits under the division symbol. Write a zero in the quotient for each of those digits.

$$\begin{array}{r} 300 \\ 7\overline{)2340} \\ \underline{21} \end{array}$$

1 Divide 7 into 23.
The largest whole number is 3.
2 Write a zero for each digit in the rest of the dividend.

The exact answer will always be **more than** your estimate.

PRACTICE

Estimate an answer for each division problem.

1 $2\overline{)145}$

2 $3\overline{)425}$

3 $8\overline{)816}$

4 $5\overline{)357}$

5 $6\overline{)3691}$

6 $4\overline{)1256}$

Here are some addition, subtraction, multiplication, and division problems. First, write an estimate for each problem. Then use a circle to indicate whether the exact answer or the estimate is more. You may want to review pages 25, 39, and 52.

7 $\begin{array}{r} 125 \\ +\ 203 \\ \hline \end{array}$

Which is more?
Estimate
Exact answer

8 $\begin{array}{r} 21 \\ \times\ 2 \\ \hline \end{array}$

Which is more?
Estimate
Exact answer

9 $\begin{array}{r} 51 \\ +\ 72 \\ \hline \end{array}$

Which is more?
Estimate
Exact answer

10 $4\overline{)203}$

Which is more?
Estimate
Exact answer

11 $\begin{array}{r} 71 \\ -\ 39 \\ \hline \end{array}$

Which is more?
Estimate
Exact answer

12 $7\overline{)714}$

Which is more?
Estimate
Exact answer

13 $\begin{array}{r} 158 \\ -\ 72 \\ \hline \end{array}$

Which is more?
Estimate
Exact answer

14 $\begin{array}{r} 319 \\ \times\ \ 2 \\ \hline \end{array}$

Which is more?
Estimate
Exact answer

15 $3\overline{)140}$

Which is more?
Estimate
Exact answer

Solving Word Problems

Set up each problem and solve it. Not all of these are division problems.

Use this information to answer questions 1 through 4.

Fred takes home $500 a week. He works Monday through Friday.

1 How much does Fred take home in one year (52 weeks)?

2 How much does Fred take home per workday?

3 Last week Fred got a bonus of $125. How much did he take home last week?

4 Fred has been offered another job that has a take-home pay of $612 per week. If he accepts the new job, how much more will he take home in pay each week?

Use this information to answer questions 5 through 8.

Anita needs 8 yards of sidewalk.
Company A will charge $320.
Company B will charge $295.
Company C will charge $30 per yard.

5 How much more will company A charge than Company B?

6 How much will Company C charge for all 8 yards?

7 How much is Company A charging per yard?

8 Anita decides to hire Company B. She gives them $30 as a down payment. How much does she have left to pay?

Solving Two-Step Word Problems

Circle the letter for the choice that correctly describes how to solve each problem.

1 Manuel bought a CD player for $250. He put $25 down. He will pay the rest in 12 equal payments. **How much will each payment be?**

- **A** Add $250 and $25. Then divide by 12.
- **B** Subtract $25 from $250. Then divide by 12.
- **C** Multiply $25 by 12. Then subtract $250.

2 Trenton bought a vase for $18 plus 74¢ tax. He paid with a $20 bill. **How much change should he get back?**

- **F** Subtract $18 from $20. Then add 74¢.
- **G** Add $18 dollars and 74¢. Then subtract the sum from $20.
- **H** Subtract 74¢ from $18. Then subtract the difference from $20.

3 Mary and Sheila split their profits evenly. Last week they earned $375 on one job and $425 on another. **How much will each woman get?**

- **A** Add $375 and $425. Then divide the sum by 2.
- **B** Divide $375 by 2. Then add $425.
- **C** Subtract $375 from $425. Then divide by 2.

4 Three months ago, Gordon weighed 210 pounds. Now he weighs 195 pounds. **About how much did Gordon lose per month?**

- **F** Divide 210 by 3. Then divide 195 by 3.
- **G** Add 210 and 195. Then divide the sum by 3.
- **H** Subtract 195 from 210. Then divide the difference by 3.

On the blank lines, describe two steps that will solve the problem. ***You do not have to solve the problems.***

5 Hank took $60 out of the bank. He spent $7 on lunch and $35 on groceries. **How much money does he have left?**

To solve, ______________________.

Then ______________________.

6 Darlene sells pillows. She bought 6 yards of cloth and used 3 yards to make one large pillow. She will divide the rest into three equal parts to make smaller pillows. **How much fabric does she have for each smaller pillow?**

To solve, ______________________.

Then ______________________.

Solve the following two-step problems.

7 Ray earns $1200 a month. His wife earns $1800 a month. **How much do they make per year?**

8 Francesca bought a book for 15 dollars plus 1 dollar tax. She paid with a 20-dollar bill. **How much change should she get back?**

9 Ronnie and Glenn want to spend $500 for a television set. Ronnie has $125. Glenn has $175. **How much more do they need?**

10 Jessica is taking her three children to the movies. She has $25 to spend. The tickets cost $19. **If she splits the remaining money evenly among her children, how much will each child get?**

11 Jeff picks apples. He is paid $50 a day, plus $3 for every bushel of apples. Today he picked 8 bushels of apples. **How much money did Jeff make today?**

Division Skills Practice

Circle the letter for the best answer to each problem. ***Make sure your answer seems reasonable.***

1 $40 \div 4 =$ ____

A 20
B 10
C 11
D 36
E None of these

2 $20 \div 5 =$ ____

F 25
G 5
H 15
J 4
K None of these

3 $98 \div 7 =$ ____

A 24
B 11 r 1
C 12 r 4
D 9 r 5
E None of these

4 $16 \div 3 =$ ____

F 5
G 6 r 1
H 6
J 5 r 1
K None of these

5 $2\overline{)186}$

A 34
B 42
C 43
D 63
E None of these

6 $4\overline{)99}$

F 80
G 100
H 5
J 10
K None of these

Use this information to do Numbers 7 through 9.

Clark is shopping for frozen orange juice. There are three brands.
Brand A: 40¢ for 8 ounces
Brand B: 64¢ for 16 ounces
Brand C: $1.60 for 32 ounces

7 Clark has $5.00. How much change will he get back if he buys 3 cans of brand C? (Do not include tax.)

A 20¢
B 70¢
C 60¢
D 50¢

8 How much juice would Clark get if he bought two cans of brand A?

F 4 ounces
G 10 ounces
H 8 ounces
J 16 ounces

9 To find out how much Brand A costs per ounce, what should Clark do?

A Add.
B Subtract.
C Multiply.
D Divide.

10 $5 \div 2 =$ ____

F 10
G 25
H 5
J 48
K None of these

11 $28 \div 4 =$ ____

A 6
B 9
C 7
D 8
E None of these

12 $68 \div 1 =$ ____

F 68
G 7
H 0
J 34
K None of these

13 $2\overline{)42}$

A 22
B 20
C 12
D 21
E None of these

14 $8\overline{)64}$

F 8
G 7
H 6
J 9
K None of these

15 $19 \div 9 =$ ____

A 2
B 2 r 2
C $2\frac{1}{2}$
D 2 r 1
E None of these

16 $2\overline{)36}$

F 13
G 19
H 17
J 18
K None of these

17 Which of these, if any, has the same value as $67 \div 5$?

A $5 \div 67$
B the product of 67 and 5
C $5\overline{)67}$
D $67\overline{)5}$
E None of these

18 Which of these number sentences, if any, is false?

F $76 \div 1 = 76$
G $63 \div 63 = 1$
H $0 \div 1 = 0$
J $66 \div 6 = 11$
K None of these

19 What sign, if any, goes in the box to make the second number sentence true?

$116 \div 29 = 4$

$29 \square 4 = 116$

A $\div$
B $\times$
C $+$
D $-$
E No sign can make the second sentence true.

Reading a Table

One way to organize numbers and information is in a **table.** The information is organized in **rows** and **columns.**

Each column has a heading. The heading tells what is described in the column.

Work Crews

Crew	Size	Current Job
A	3	Sheldon Home
B	4	Waverly Home
C	2	New Library

Read each row from left to right.

Read each column from top to bottom.

PRACTICE

Use the Work Crews table above to answer each question.

1 What is this table about?

A Libraries
B Different types of jobs
C The sizes of different homes
D Workers, crews, and jobs

2 What is this part of the table called?

C	2	New Library

F a column
G a row
H a heading
J the title

3 What does this part of the table tell you?

3
4
2

A the size of each crew
B who is on each crew
C which crew is best
D which jobs are being done

4 What does the third column in the table describe?

F crew C
G the Waverly Home
H the size of each crew
J what job each crew is working on

Suppose you want to know the cost of a large coffee drink with steamed milk. Look across the row for "steamed milk" and look down the column for "large."

Look down this column.

Specialty Drinks

Coffee Drink	Small	Medium	Large
Flavored Coffee	$0.85	$1.05	$1.25
Latte	$1.45	$1.75	$2.05
Steamed Milk	$1.65	$1.95	$2.25
Mocha	$2.35	$2.70	$3.05

Look across this row.

The cost of a large coffee drink with steamed milk is $2.25.

PRACTICE

Use the Specialty Drinks table above to answer these questions.

5 What is the cost of a large flavored coffee? ________________

6 How much is a medium latte? ________________

7 How much is a large mocha? ________________

8 Maureen wants to spend no more than $1.00. Which drink can she buy?

____

9 Which drink costs $1.95? ________________

10 Which drink costs $1.45? ________________

Using Numbers in a Table

It can be easy to compare numbers in a table. Suppose you want to know which month had the most reported cases of vandalism.

Winter Crimes

	December	January	February
Theft	28	22	20
Burglary	10	9	5
Vandalism	16	15	18

1 Look at the row for vandalism.
2 Find the largest number in that row.
3 Find the heading for that column.

PRACTICE

Use the Winter Crimes table above to answer these questions.

1 Which month reported the most burglaries? ____________

2 What was the most commonly reported crime in January? ____________

3 Which month had the most crime reports overall? ____________

4 Which month had the *least* number of vandalism reports? ____________

5 **Circle the letter of the choice that best describes burglary reports in February.**

A It was the most common crime.

B It was the least common crime.

C It was more commonly reported than vandalism, but less commonly reported than theft.

Comparing Numbers in a Table

There are several ways to compare two numbers in a table.

- You can subtract to find the **difference** between two numbers.
- You can divide to see **how many times larger** one number is than another.
- You can form a fraction to find **what part** one number is of another.

Local Bike Trails

Trail	Old Mill	Sauk	Ruby Creek	Hines Park	Interstate 80	Lakefront
Length (in miles)	5	10	15	20	30	40

Question 1: The Ruby Creek trail is ? times longer than the Old Mill trail.
Solution Ruby Creek: 15 miles
Old Mill: 5 miles
$15 \div 5 = 3$
The Ruby Creek trail is 3 times longer than the Old Mill trail.

Question 2: The Old Mill trail is ? times the length of Sauk trail.
Solution Old Mill trail: 5 miles
Sauk trail: 10 miles
$\frac{5}{10}$
The Old Mill trail is five-tenths of the length of Sauk trail.

PRACTICE

Use the table above to answer these questions.

1 Circle the **longer** trail:
Ruby Creek trail or Sauk trail.

2 Which is **shorter,** the Hines Park trail or the Old Mill trail?

3 How many miles longer is the lakefront trail than the Interstate 80 trail? ____________

4 Ricki is choosing a bicycle trip on either the lakefront trail or the Hines Park Trail. How many miles farther can she ride if she takes the lakefront trail?

____

5 The Hines Park trail is ______ times longer than the Sauk trail.

6 The Interstate 80 trail is ? times longer than the Old Mill trail.

7 The Old Mill trail is ? times the length of the Ruby Creek trail. *(Write a fraction.)*

8 Sauk trail is ? times the length of the lakefront trail. *(Write a fraction.)*

9 The Ruby Creek trail is what fraction of the length of the Hines Park trail? ____________

10 The lakefront trail is ? times longer than the Old Mill trail.

11 The Old Mill trail is ? the length of the Hines Park trail.

Using a Calendar

You can think of a calendar as a table with seven columns. Each row is seven days, or one week.

November

Sun.	Mon.	Tues.	Wed.	Thurs.	Fri.	Sat.
						1
2	3	4	5	6	7	8
9	10	11	12	13	14	15
16	17	18	19	20	21	22
23	24	25	26	27	28	29
30						

Veterans Day → 11

Thanksgiving Day → 27

PRACTICE

Use the calendar above to answer these questions.

1 What day of the week is November 12? ________________

2 What day of the week is the last day of November? ________________

3 How many days are between Veterans Day and Thanksgiving Day? ________________

4 What date is the third Thursday of this month? ________________

5 How many Tuesdays are in this month? ________________

6 What date is Veterans Day? ________________

Using a Price List

Menus and price lists are tables. Do not be confused when you are asked to look at several prices all at once! Just find each price, one at a time. Write them down as you go. Then do any figuring.

Ben's Deli	Price per pound (tax included)
Smoked Turkey Breast	$5.20
Baked Ham	$4.00
Swiss Cheese	$4.00
Seafood Salad	$6.50
Pasta Salad	$5.00
Marcaroni and Cheese	$2.50

PRACTICE

Use the menu above for Ben's Deli to answer these questions. *Hint:* If you must buy two or more things, you *add.* If you must buy several of one thing, you *multiply.*

1 How much is a pound of smoked turkey breast and a pound of Swiss cheese?

____

2 Shirley bought a pound of pasta salad and a pound of baked ham. She also spent $0.80 on plastic forks. How much did she spend?

____

3 How much more per pound is seafood salad than pasta salad?

____

4 How much does it cost to buy 3 pounds of baked ham?

____

5 Franco bought a pound of Swiss cheese, a pound of baked ham, and a pound of pasta salad. He spent $1.25 on plastic forks. How much did he spend?

____

6 Ella only has $20.00. About how many pounds of seafood salad can she buy?

____

7 Baked ham is _?_ times the price of pasta salad. *(Write a fraction.)*

8 What dish is one-half the price of pasta salad?

____

Graphs

A graph represents information. Graphs are becoming more popular all the time. You find them in most newspapers, magazines, and reports.

Here are four types of graphs.

A **pictograph** uses pictures or symbols to represent numbers.

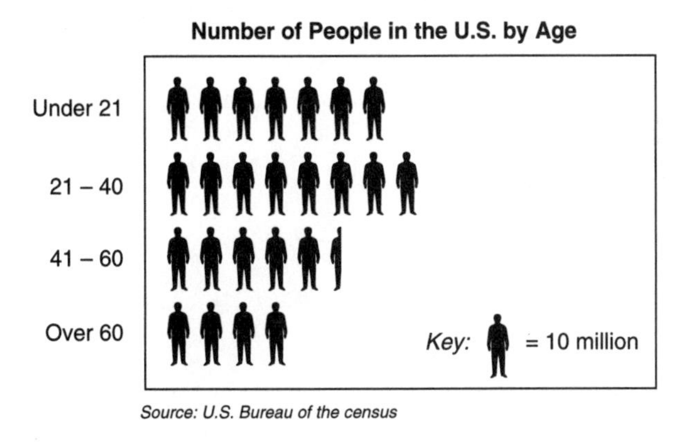

1 How many categories of ages are there?

A **circle graph** divides a circle into slices or wedges to show parts of a whole.

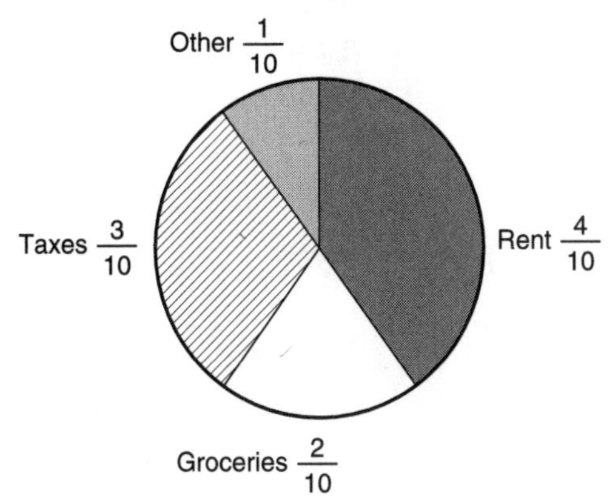

2 How many categories of expenses are shown?

A **bar graph** uses thick bars to represent numbers. The bars can be vertical (up-and-down) or horizontal (left-to-right).

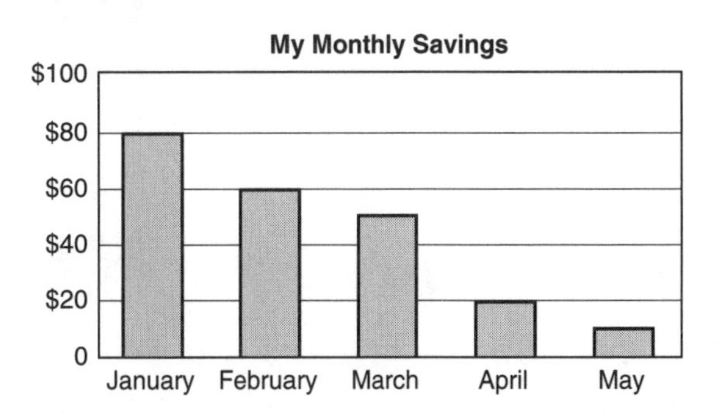

3 What month shows the largest savings?

A **line graph** uses dots to show numbers. Lines connect dots to show rising and falling values.

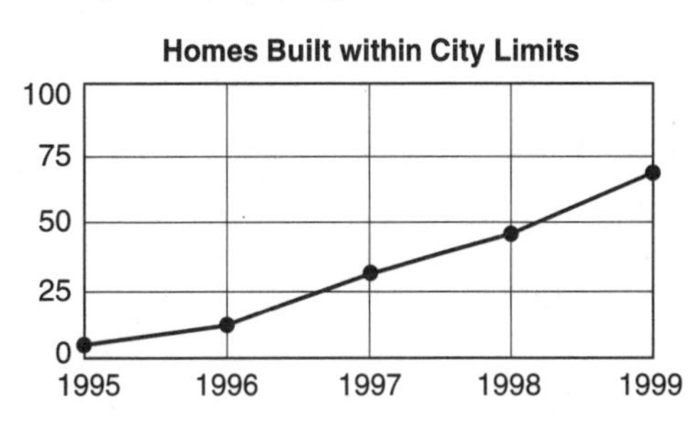

4 How many years are shown?

5 In what year were the most homes built?

Reading a Pictograph

In a pictograph, each picture stands for a particular number. In the graph below, the **key** tells you how many people are shown by each figure. (*If there is no key, each picture on the graph stands for* 1.)

To read a pictograph, count the number of pictures next to a label. Then multiply that number by the value shown in the key.

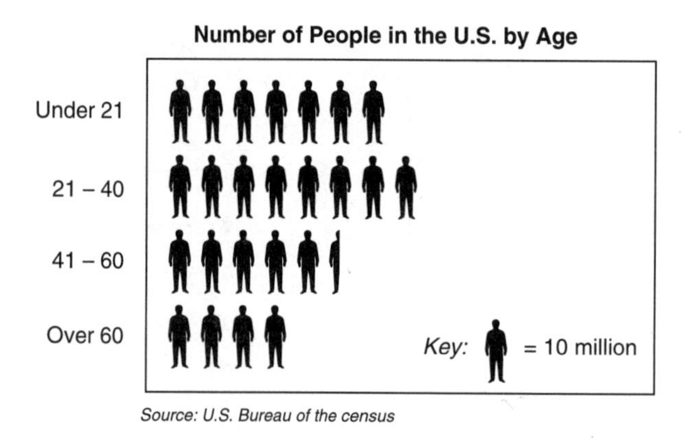

Question: About how many Americans are over 60 years old?
Solution:

The key says that one figure represents 10 million people.
There are 4 figures in the "Over 60" group.
4×10 million people = 40 million people

PRACTICE

Use the pictograph above to answer these questions.

1 Each figure on this chart stands for _?_ people. ________

2 The biggest age group shown in the graph is _?_ ________

3 About how many people in the United States are under 21?

____

4 There are about ____________ times as many Americans 21–40 as there are over 60.

5 There is a picture of one-half a figure shown at the end of one line. What does that represent?

____

6 Circle the larger group:

Americans under 21   Americans 41–60

7 The total number of Americans is about 250 million people. About what fraction of that population is over 60?

____

Reading a Bar Graph

A bar graph uses thick lines, or bars, to represent numbers. The longer the bar, the bigger the number.

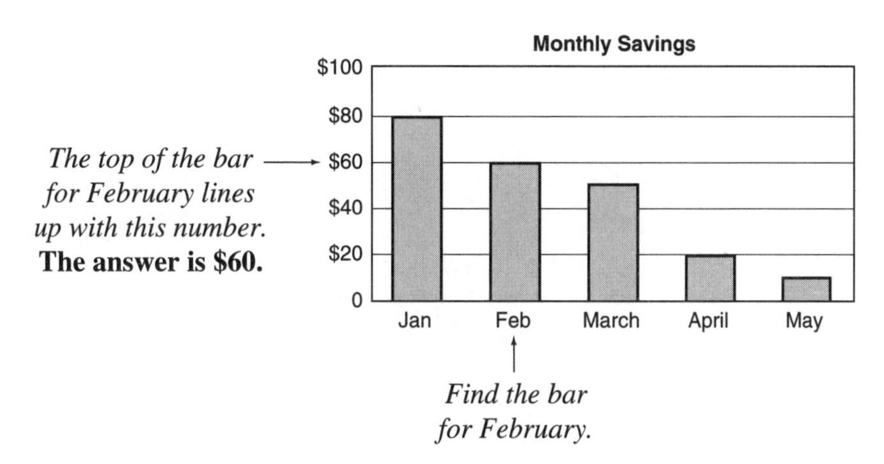

Question: How much money did this person save in February?
Solution:
1 Find the bar for February.
2 The top of the February bar represents $60.
3 The person saved $60 in February.

PRACTICE

The bar graph above shows Sophie's savings. Use the graph to answer these questions.

1 How much money did Sophie save in January?

____

2 How much money did she save in April?

____

3 The bar for May ends halfway between $0 and $20. How much money did Sophie save in May?

____

4 How much money did Sophie save altogether in January and February?

____

5 What is the difference between the largest amount and the smallest amount that Sophie saved in a month?

____

6 Sophie saved about one-third as much in April as she saved in which month?

____

7 Sophie's February savings were about ? of her January savings. *(Write a fraction.)*

____

Reading a Circle Graph

A circle graph shows how a total is divided into parts. It uses sections, like a pie. Each section stands for a different part, or fraction, of the total. The larger the section, the larger the fraction.

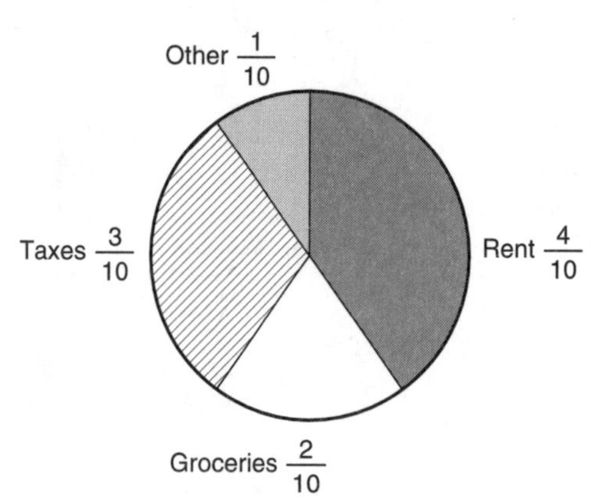

PRACTICE

The graph above shows how Eric spends his money each month. Use the graph to answer the questions below.

1 What fraction of the money does Eric spend on groceries?

____

2 The biggest fraction of Eric's spending is on which category?

F rent
G groceries
H taxes
J other

3 Together, taxes and groceries make up about what fraction of Eric's spending?

A $\frac{1}{10}$
B $\frac{2}{10}$
C $\frac{3}{10}$
D $\frac{5}{10}$

4 Which category would include money Eric spent on clothing?

F rent
G groceries
H taxes
J other

5 How much is Eric's monthly rent?

A \$300
B \$130
C \$333
D It is impossible to tell from the graph.

6 Eric spends twice as much money on rent as on which category?

F taxes
G groceries
H other
J taxes and groceries

Finding an Average

Look again at the lengths of these local bike trails. Suppose you want one number to represent a "typical" trail length. One way to find a typical value is to find the **average** (or **mean**).

Local Bike Trails

Trail	Old Mill	Sauk	Ruby Creek	Hines Park	Interstate 80	Lakefront
Length (mi)	5	10	15	20	30	40

To find an average, *add* all the numbers. Then *divide* that sum by the number of values.

Question: What is the average length of the local bike trails?

Solution:

1 Add all the numbers.

$$\begin{array}{r} 5 \\ 10 \\ 15 \\ 20 \\ 30 \\ +\ 40 \\ \hline 120 \end{array}$$

The sum is 120.

2 There are 6 numbers in the list, so divide the sum by 6.

$$6\overline{)120}\text{ = } 20$$

PRACTICE

1 **Find each runner's average time. Enter each average time in the shaded column.**

Time in Trial Races (Minutes)

	1	2	3	4	5	6	Average Time (in minutes)
Jackie	9	10	11	10	10		
Jordan	10	10	12	12	10	12	
Rene	12	17	11	9	11		
Jamie	12	10	10	10	13		

Use common sense estimation to answer questions 2 and 3.
***Hint:* the average will be near the middle of a set of numbers.**

2 Gwen completed five speed tests in typing class. Her times, in minutes, were 2, 3, 2, 5, and 4. Which is the best estimate of her average time?

F 2 **H** 5
G 3 **J** 6

3 On the state drivers license test, Jean and her friends got the following scores: 65, 78, 63, 78, 83, 42, 53. About what was their average score?

A 42 **C** 64
B 51 **D** 81

Data Interpretation Skills Practice

Celeste works in a biology laboratory. Her tasks include observing insects and describing what they do. This table shows how many she observed in the lab last week. **Study the table. Then do questions 1 through 8 on pages 84 and 85.**

Insects in the Lab

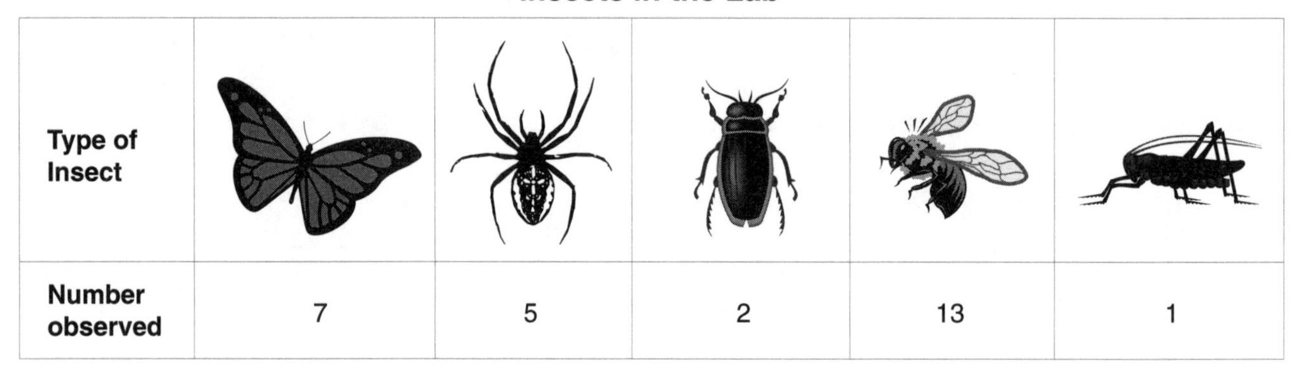

Type of Insect					
Number observed	7	5	2	13	1

1 What does the fourth column from the right show?

A how many spiders Celeste observed
B how many beetles Celeste observed
C the number of times Celeste observed each type of insect
D all the different types of insects Celeste observed

2 Celeste observed exactly 5 of which bug?

F 

H 

G 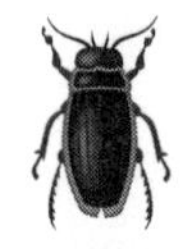

J 

3 How many butterflies did Celeste observe?

A 10 **C** 1
B 7 **D** 3

4 How many more spiders than beetles did Celeste observe?

F 7
G 2
H 3
J 6

5 This week, Celeste observed twice as many beetles as last week. How many beetles did she observe this week?

A 1 **C** 4
B 3 **D** 6

6 Which bug did she observe least often?

F butterflies **H** spiders
G bees **J** crickets

7 How many insects did Celeste observe in all?

A 18 **C** 20
B 15 **D** 28

8 Celeste observed twice as many beetles as

F butterflies **H** bees
G spiders **J** crickets

The chart below shows the number of bees Celeste watched since she started her job. **Use the chart to answer questions 9 through 14.**

Number of Bees Watched Each Week

Week	1	2	3	4	5	6	7	8
Bees	15	11	11	10	10	8	8	7

9 How many weeks has Celeste been watching bees?

A 2
B 7
C 8
D It is impossible to tell.

10 Celeste observed a total of 80 bees. What fraction of that number did she observe in the third week?

F $\frac{7}{80}$ **H** $\frac{11}{80}$
G $\frac{80}{11}$ **J** $\frac{8}{80}$

11 In the first two weeks combined, Celeste observed a total of how many bees?

A 15
B 4
C 16
D 26

12 Celeste observed the same number of bees in the sixth week as she did in which week?

F fifth **H** first
G eighth **J** seventh

13 What is the average number of bees Celeste observed each week?

F 13 **H** 10
G 11 **J** 8

14 The number of bees that Celeste is watching seems to be

A rising
B falling
C staying the same
D None of the above

Pre-Algebra

Finding Patterns

A key to understanding math is to see and use patterns. When you describe a pattern using letters and symbols, you are using the math ideas called **algebra.**

PRACTICE

Each box on this page contains a pattern. There may be a pattern in the number of objects, their shape, their shading, or some other feature.

Find each pattern. Then, on the blank line, draw the figure that comes next.

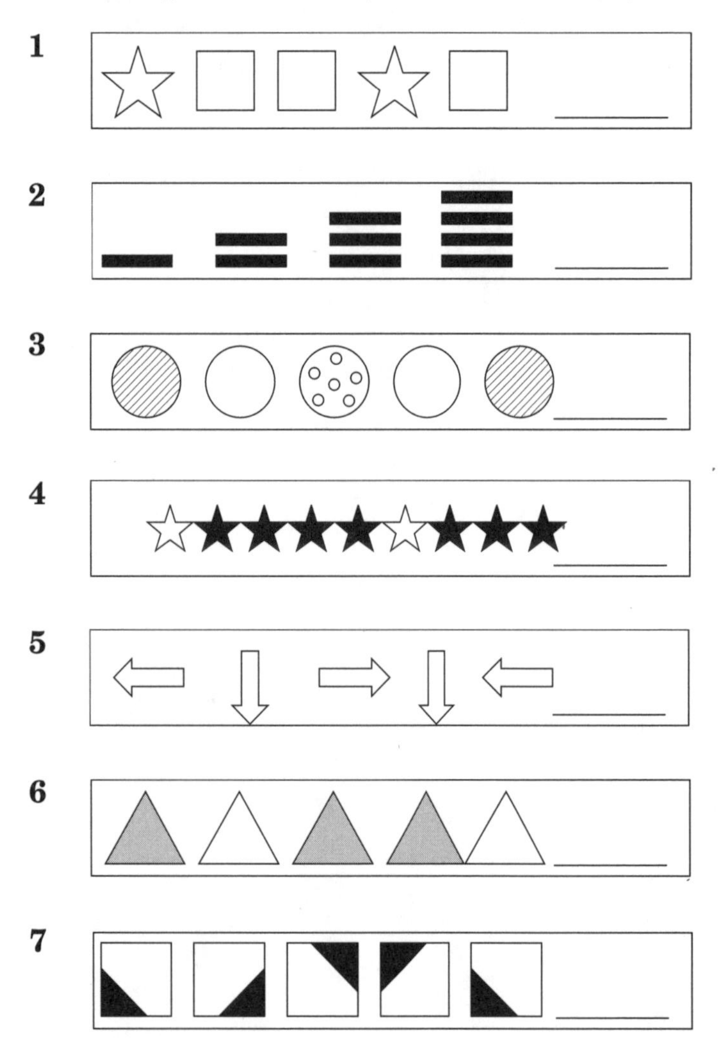

Answer each question.

8 How many lines will the fifth figure have?

A 3 **C** 5
B 4 **D** 6

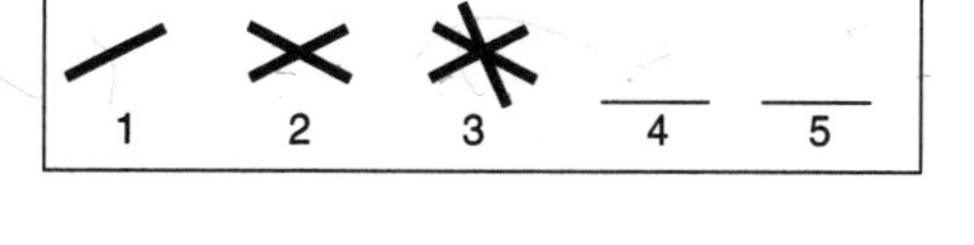

9 Will the thirteenth line be thick or thin?

____

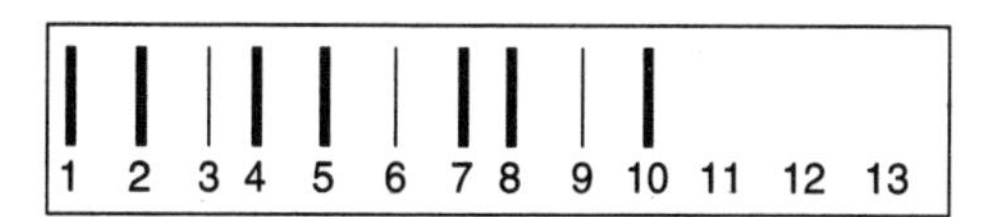

10 Which part of the ninth box will be shaded?

A the top
B the middle
C the bottom
D none of it

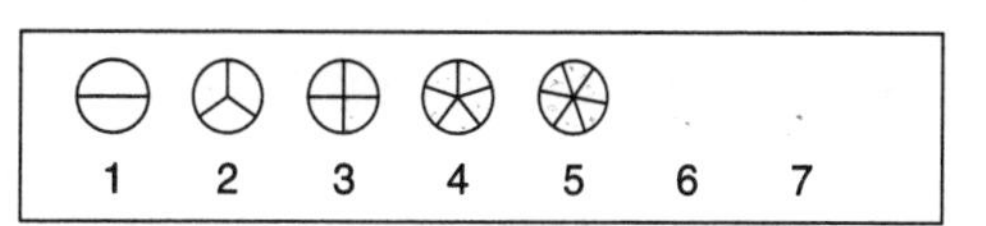

11 The seventh circle will be divided

into ________ parts.

12 Circle the letter for the figure that will be next.

A **C**

B **D**

13 The diagram below shows how a bathroom floor is being decorated. The bottom row has not been done.

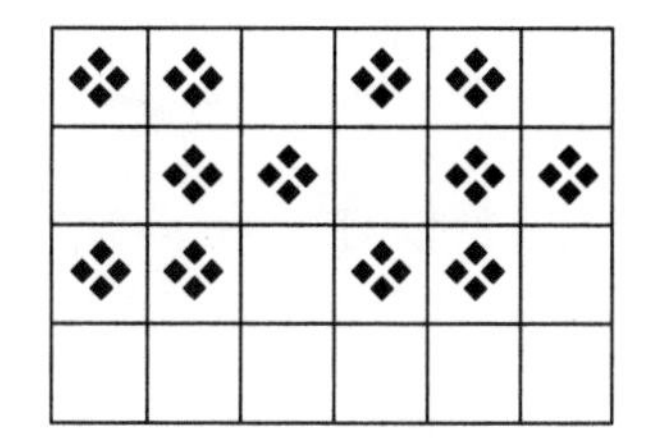

What will the bottom row look like?

A

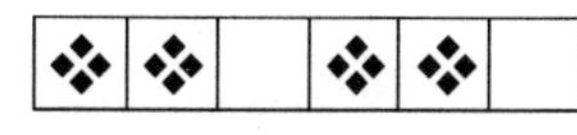

B

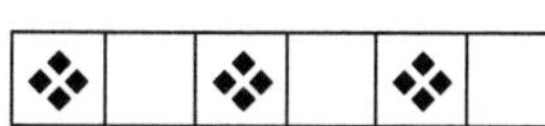

C

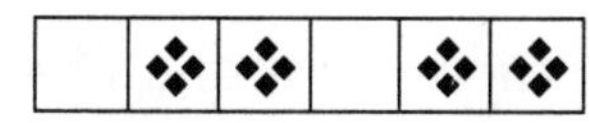

D

Finding Number Patterns

You can create a pattern with numbers as well as with shapes. Here are some number pattern problems. Answer each question by filling in the blanks.

PRACTICE

1 Starting with two, count by twos:

2, 4, 6, ____, ____, ____, ____

2 Starting with ten, count by fives:

10, 15, 20, ____, ____, ____

3 Starting with three, count by threes:

3, 6, 9, ____, ____, ____

4 Here is a number pattern:

4, 8, 12, 16, 20, 24, . . .

This pattern shows counting by

____, starting with ____.

5 Here is a number pattern:

1, 3, 5, 7, 9, 11, . . .

In this pattern, each new number

is the previous number plus ____.

6 Here is a number pattern:
25, 20, 15, 10, 5, 0
In this pattern, each new number is the previous number minus

____.

7 Here is a number pattern:
2, 4, 8, 16, 32, . . .
In this pattern, each new number is the previous number times

____.

8 Describe this number pattern:
0, 6, 12, 18, 24, . . .

____

9 Describe this number pattern:
1, 5, 9, 13, 17, 21, 25, . . .

____

10 Describe this number pattern:
90, 80, 70, 60, 50, . . .

____

11 Fill in the missing number in this pattern: 2, 5, 8, 11, ____, 17, 20, 23, 26.

12 Fill in the missing numbers in this pattern: 6, 10, 14, ____, 22, 26, ____, 34.

13 Fill in the missing number in this pattern: 1, 3, 9, ____, 81, 243.
(*Hint:* It is a multiplication pattern.)

Patterns in Multiplication and Division Facts

You can find patterns when you multiply and divide. Number facts can be easier to remember if you can use these patterns.

PRACTICE

The table below shows some numbers made when you multiply by 2. These are called *multiples of 2*.

$2 \times 1 = 2$
$2 \times 2 = 4$
$2 \times 3 = 6$
$2 \times 4 = 8$
$2 \times 5 = 10$
$2 \times 6 = 12$
$2 \times 7 = 14$
$2 \times 8 = 16$
$2 \times 9 = 18$

1 To find the multiples of two, start

with ____ and count by ____.

2 The multiples of 2 are also called

the ____ numbers.

Here are the first ten numbers that can be evenly divide by five:

5, 10, 15, 20, 25, 30, 35, 40, 45, 50.

3 If a number ends with ____ or

____, it can be evenly divided by 5.

4 According to problem 3, which of the following numbers can be evenly divided by 5?

A 102 **C** 552
B 130 **D** 503

5 Here are the first eight multiples of 9. Look at the ones digits. As you move down the column, the ones digits decrease by one.

$9 \times 1 = 9$
$9 \times 2 = 18$
$9 \times 3 = 27$
$9 \times 4 = 36$
$9 \times 5 = 45$
$9 \times 6 = 54$
$9 \times 7 = 63$
$9 \times 8 = 72$

As you move down the column, what pattern appears in the *tens* digits?

____

6 Use the pattern in problem 5 to find the next multiple of nine.

____

7 When you multiply a number by 10, the ones digit in the product is

always ____. The other digits are always the same as

____

8 Use the pattern in problem 7 to find 65 times 10. ______

9 All the multiples of 10 are also

multiples of ____. (*Hint:* Look at the lists on this page.)

Patterns in Graphs and Tables

Often, there are patterns or trends in tables and graphs. If you can identify the pattern, you can predict what value will come next.

PRACTICE

Specialty Drinks

Coffee Drink	Small	Medium	Large	Extra Large
Flavored Coffee	$0.85	$1.05	$1.25	
Latte	$1.45	$1.75	$2.05	
Steamed Milk	$1.65	$1.95	$2.25	
Mocha	$2.35	$2.70	$3.05	

Use the table above to answer questions 1 through 4.

1 Read the row for "Latte" from left to right. By how much does the price increase?

____

2 Read the row for "Flavored Coffee" from left to right. By how much does the price increase?

____

3 If the pattern of prices continues in the "Flavored Coffee" row, what would an "Extra Large Flavored Coffee" cost?

____

4 If the pattern of prices continues in the "Mocha" row, what would an "Extra Large Mocha" cost?

____

Use this graph to answer questions 5 and 6.

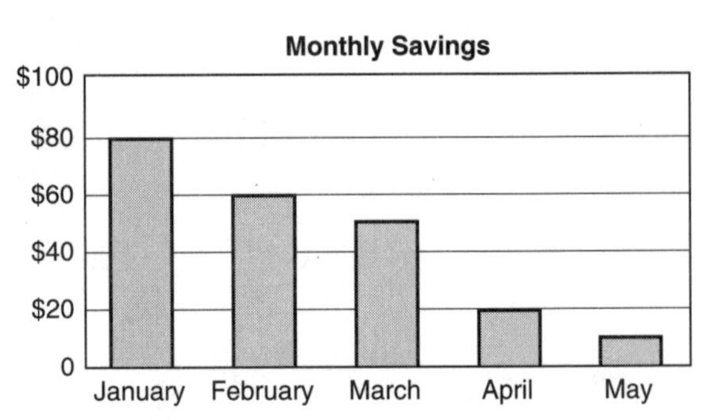

5 Every month, does the amount saved increase or decrease?

____

6 If this trend continues, about how much money will this person save in June?

A about $30
B about $50
C about $15
D less than $10

Patterns in Number Sentences

The number sentences below are incomplete. Use what you know about number facts to answer each question.

PRACTICE

Circle the correct sign.

1 + − × ÷
Circle the sign above that belongs in all three number sentences.
2 ☐ 2 = 4
10 ☐ 2 = 12
8 ☐ 3 = 11

2 + − × ÷
Circle the sign above that belongs in all three number sentences.
10 ☐ 3 = 7
8 ☐ 4 = 4
9 ☐ 1 = 8

3 + − × ÷
Circle the sign above that belongs in all three number sentences.
2 ☐ 3 = 5
3 ☐ 5 = 8
1 ☐ 3 = 4

4 + − × ÷
Circle the sign above that belongs in all three number sentences.
5 ☐ 1 = 6
42 ☐ 3 = 45
21 ☐ 21 = 42

5 + − × ÷
Circle the sign above that belongs in all three number sentences.
32 ☐ 10 = 22
9 ☐ 4 = 5
15 ☐ 9 = 6

Fill in each blank with a number.

6 2 + ☐ = 7
10 + ☐ = 15
0 + ☐ = 5
What number makes each of these sentences true? _______

7 3 + ☐ = 4
5 + ☐ = 6
9 + ☐ = 10
What number makes each of these sentences true? _______

8 5 − ☐ = 2
8 − ☐ = 5
12 − ☐ = 9
What number makes each of these sentences true? _______

9 9 − ☐ = 4
5 + ☐ = 10
5 ÷ ☐ = 1
What number makes each of these sentences true? _______

10 3 × ☐ = 0
4 + ☐ = 4
5 − ☐ = 5
What number makes each of these sentences true? _______

For each problem, look at the three number sentences. What goes into each box to make all three sentences true?

Sample

5 ☐ = 7
2 ☐ = 4
1 ☐ = 3

Add 2.

11
5 ☐ = 11
1 ☐ = 7
10 ☐ = 16

____

12
7 ☐ = 14
2 ☐ = 9
1 ☐ = 8

____

13
8 ☐ = 7
3 ☐ = 2
4 ☐ = 3

____

14
6 ☐ = 8
2 ☐ = 4
1 ☐ = 3

____

For each problem, tell what rule will change each "In" number to the corresponding "Out" number.

Sample

In	3	4	5	7
Out	1	2	3	5

Subtract 2.

15

In	7	9	6	10
Out	1	3	0	4

____

16

In	1	6	10	5
Out	6	11	15	10

____

17

In	2	3	8	5
Out	4	6	16	10

____

18

In	15	20	30	12
Out	5	10	20	2

____

Writing Letters and Symbols for Words

An algebra problem often uses letters to take the place of numbers. These letters are called **variables** or **unknowns.** A letter can represent a single number or it can represent many numbers.

In this table, each "Out Number" can be described as "In Number − 2."

In	3	4	5	7
Out	1	2	3	5

There is another way to describe the numbers. If we use the letter n to describe each of the "In" numbers, then each of the "Out" numbers is $n - 2$.

Algebra also uses the symbols for addition (+), subtraction (–), multiplication (×), and division (÷). For example, the phrase "*a number increased by* 5" can be represented as $n + 5$.

PRACTICE

Write an algebraic expression for each phrase. Let the letter *n* stand for the unknown. *(If you have trouble, review page 53.)*

1 a number plus twelve

____

2 a number minus three

____

3 the sum of a number and four

____

4 one more than a number

____

5 three less than a number

____

6 two times a number

____

7 the total of five and a number

____

8 a number divided by six

____

9 three divided by a number

____

10 a number decreased by thirty-two

____

11 seven multiplied by a number

____

12 12 increased by a number

____

Solving a word problem is an important use of algebra letters and symbols. The first step is to look for signal words for addition, subtraction, multiplication, and division.

Write an algebraic expression to describe each of these situations.

Sample
Jack makes w dollars an hour. Starting next month, he will get $2.00 more per hour. What will his new hourly wage be?
Answer: $w + 2$ dollars

13 There are s students enrolled in Linda's math class. Tonight, two of them are absent. How many students came to class tonight?

_______ students

14 Geraldo had x dollars in his wallet. He gave $15.00 to his son. How many dollars does Geraldo have now?

_______ dollars

15 Elena earns F dollars per week at her first job. She makes S dollars per week at her second job. What is her total weekly pay?

_______ dollars

16 A tablecloth is c feet long. It will be made 2 feet longer to hang properly. What will be the final length of the tablecloth?

_______ feet

17 Rick wants to plant b bushes. Each bush needs 4 feet of space. How many feet of space will his bushes need altogether?

_______ feet

18 Trish has 12 students. She does not know how many students Claudio has. (Let C stand for that number of students.) Write an expression for the number of students they have altogether.

_______ students

19 Clint has 96 bricks. He will have to give some of them to his brother. (Let b stand for that number of bricks.) What expression shows how many bricks Clint will have left?

_______ bricks

20 Ken must bake 3 cakes for his daughter's science club. He must also bake cakes for the Girl Scouts. (Let g stand for that number of cakes.) Write an expression that shows how many cakes he must bake altogether.

_______ cakes

Writing Equations

To solve a word problem using algebra letters and symbols, you must write an **equation.** An equation is an expression such as $x + 7 = 10$. It has an "equal" sign, which says that there is the same value on each side of that sign.

Example 1: Armen is 48 years old. How old was he 5 years ago?

What operation should I use?
The phrase "5 years ago" is a signal for subtraction.
What number(s) do I use?
Armen's present age (48) is the total, so you must subtract 5 years *from* that.
Equation: $48 - 5 = x$

Example 2: Terrence just bought 3 new wrenches. He now has a total of 15 wrenches. How many wrenches did Terrence have to begin with?
What operation should I use?
The word "total" is a signal for addition.
What number(s) do I use?
The total 15 is given, so you must add the other two numbers together to get 15.
Equation: $x + 3 = 15$

PRACTICE

For each problem, write an equation that describes the problem. Use the letter *x* for each unknown. (*Hint:* You subtract *from* a total. To get a total, you *add.*)

1 Clyde had 3 cars. He just bought 2 more. How many cars does Clyde have now? ____________

2 Larry has eaten 6 cookies. There are 8 cookies left in the box. How many cookies were there originally? ____________

3 Candice has 3 dogs. Her fiancée has 4 dogs. How many dogs do they have altogether? ____________

4 There is a difference of 7 years in the ages of Annelle's children. Her youngest child is 6 years old. What age is her oldest child? ____________

5 Glenda has 8 grandchildren. Five of them are girls. How many are boys? ____________

6 Rochelle had 8 presents to give away. She gave 2 of the presents to friends. then she gave her children another 2 of the presents. How many presents does she have left? ____________

Solving Equations

The equation $x + 3 = 15$ means "Add an unknown number and 3. The result is 15." To solve that equation, perform the inverse (or opposite) of the operation shown. The inverse of "add 3" is "subtract 3." So to solve the equation, subtract 3 from each side of the equation.

To solve an addition problem, you subtract.

Equation: $x + 3 = 15$

Subtract 3 from each side: $-3 \quad -3$

$x + 0 = 12$

Adding 0 to a number does not change its value.

$x = 12$

To solve a subtraction problem, you add.

$$\begin{array}{rcr} x - 3 & = & 6 \\ +3 & & +3 \\ \hline x & = & 9 \end{array}$$

The expression $5 + x$ is the same as $+5 + x$.

To get rid of the +5, subtract 5 from each side.

$$\begin{array}{rcr} 5 + x & = & 7 \\ -5 & & -5 \\ \hline x & = & 2 \end{array}$$

PRACTICE

Solve each of the following equations. Then check your answer by seeing whether or not it makes the original equation true. *Remember:* You always do the same thing to both sides of an equation.

1 $x + 2 = 7$ $x =$ ______

2 $x + 4 = 6$ $x =$ ______

3 $3 + x = 7$ $x =$ ______

4 $x - 7 = 2$ $x =$ ______

5 $8 - x = 5$ $x =$ ______

6 $x + 1 = 5$ $x =$ ______

7 $x + 2 = 5$ $x =$ ______

8 $x + 6 = 19$ $x =$ ______

9 $x - 50 = 30$ $x =$ ______

10 $21 + x = 25$ $x =$ ______

11 $16 + x = 27$ $x =$ ______

12 $x + 11 = 81$ $x =$ ______

Use algebra to solve these word problems. First, write an equation to describe each problem. Then solve your equation.

13 A number increased by nine is fifteen. What is the number?

equation ______________

solution ______________

14 When you subtract 3 from a number, the difference is 7. What is the number?

equation ______________

solution ______________

15 Ronald bought a book for 5 dollars less than the original price. He paid 10 dollars. What was the original price?

equation ______________

solution ______________

16 Maria has five coats. That is three more coats than Anna has. How many coats does Anna have?

equation ______________

solution ______________

Pre-Algebra Skills Practice

Circle the letter for the correct answer to each problem.

1 Which number goes into the box to make this number sentence true?

2 + □ = 9

A 3
B 4
C 5
D 7

2 Which sign completes this number sentence?

7 □ 4 = 3

F +
G ×
H –
J ÷

3 This table shows "In" numbers that have been changed by a rule to get "Out" numbers. Which of these could be the rule that changes the "In" numbers to the "Out" numbers?

In	1	3	4	8
Out	3	5	6	10

A Add 1.
B Add 2.
C Add 3.
D Add 5.

4 Which of these will come next in the repeating pattern?

□ ◆ ● ◆ □ ◆ ● ◆ □ __

F □

G ◆

H ●

J ■

5 Which pattern needs a circle in the blank space?

A ▮▮ ○ ▮▮ ○ ▮▮ ○ ▮ __

B □ ○ □ ○ □ ○ __

C ○ □ ○ □ ○ □ __

D ○ ▲ ○ ○ ▲ ○ ○ __

6 What number comes after 30 when you count by tens?

F 20
G 40
H 60
J 80

This is a design for a new parking lot. The shaded section is for visitor parking.

Study the diagram. Then do numbers 7 through 9.

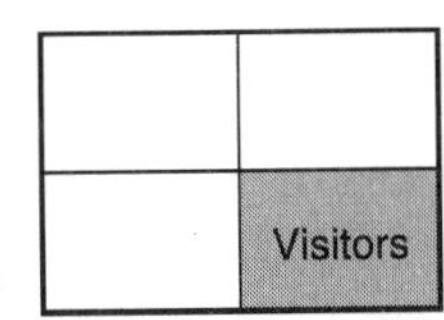

7 The parking lot will have four sections. In which number sentence is N the number of sections where visitors may park?

A $3 + N = 4$
B $3 + 1 = N$
C $1 + N = 4$
D $N - 3 = 4$

8 This diagram shows how the parking spaces will be numbered. What number will the blank space have?

103 105 107 109

F 108
G 111
H 110
J 113

9 To park in the parking lot, you must pay rent plus a deposit. In the number sentence

$\$40.00 + D = \$75.00$

D is the deposit. How much is the deposit?

A \$40.00
B \$35.00
C \$30.00
D \$75.00

Measurement

Choosing the Best Tool

There are many tools that help you measure objects. Below are some of the tools for measuring **weight, capacity, time, length, speed,** and **temperature.**

November

Sun.	Mon.	Tues.	Wed.	Thurs.	Fri.	Sat.
						1
2	3	4	5	6	7	8
9	10	11	12	13	14	15
16	17	18	19	20	21	22
23	24	25	26	27	28	20
30						

A

B

C

D

E

F

G

H

PRACTICE

On the blank line, write the letter of the best tool for making each measurement.

1 the length of a room ________

2 an amount of water for soup ________

3 how long it takes to drive to town ________

4 the size of a box ________

5 whether or not bath water is too hot ________

6 your weight ________

7 the weight of a coin ________

8 an amount of flour to make bread ________

9 days left until your birthday ________

10 how wide a flower pot is ________

Reading a Scale

A number line on a measurement tool is called a **scale.** On the scale below, the pointer rests between two labels. How do you read the measurement on the scale?

Look at the pointer and the two labels that it is between. For example, is it $\frac{3}{4}$ of the way to the higher label or is it only $\frac{1}{2}$ of the way? Often there will be little marks on the scale to help you figure this out. These are called **tick marks.**

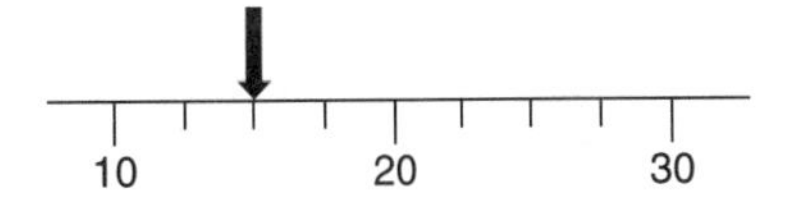

The pointer is about halfway between 10 and 20.
15 is halfway between 10 and 20.
The reading is about 15.

PRACTICE

Fill in each blank.

1

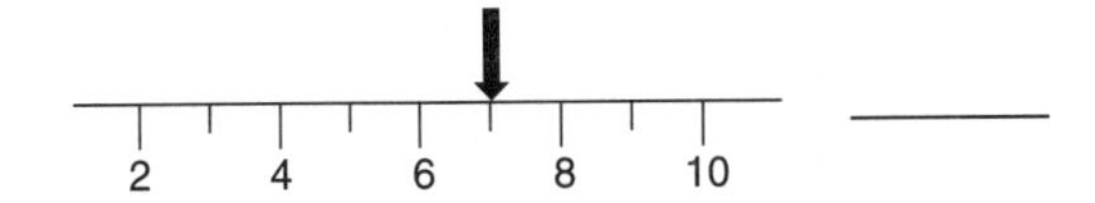

______

This pointer is about halfway between 6 and 8. What number is halfway between 6 and 8?

2

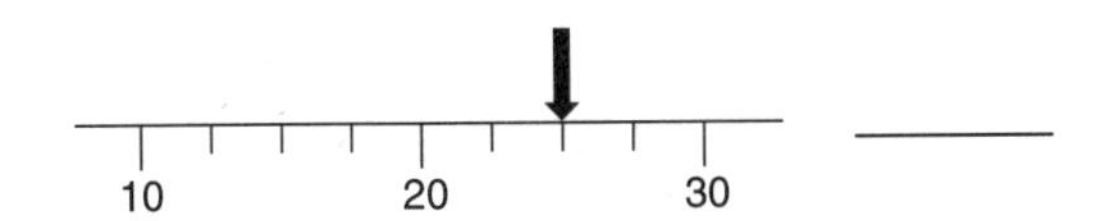

______

This pointer is about halfway between 20 and 30. What number is halfway between 20 and 30?

3

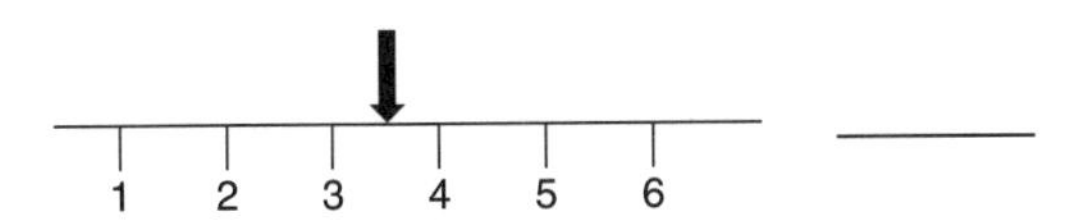

______

This pointer is halfway between 3 and 4. What number is shown on the scale?

4

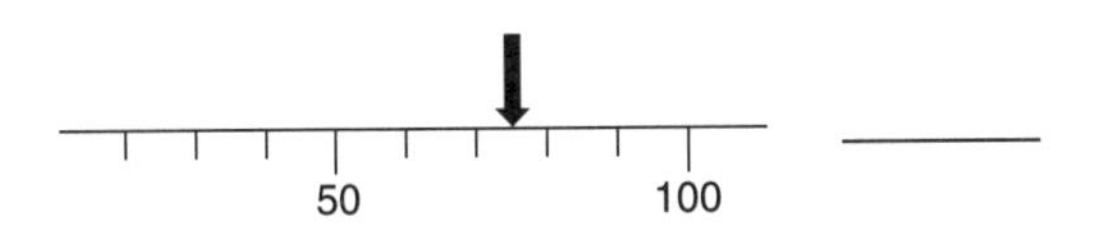

______

The pointer is halfway between 50 and 100. What number is shown on the scale?

5

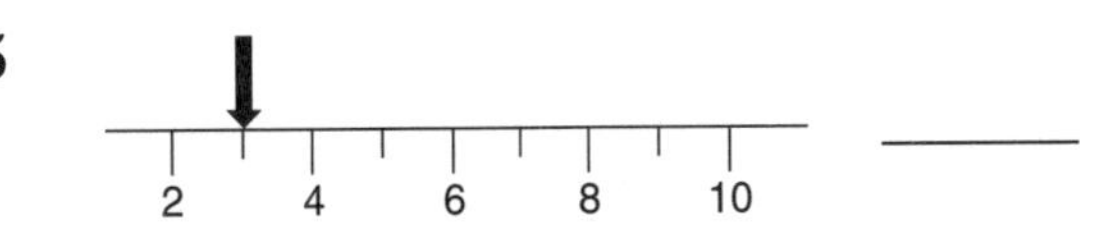

______

What number is shown on the scale?

6

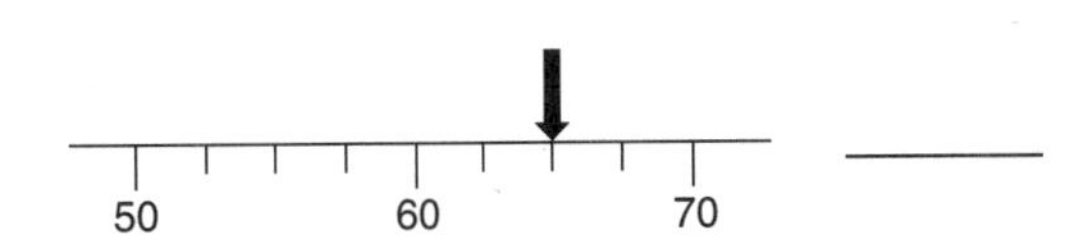

______

What number is shown on the scale?

7

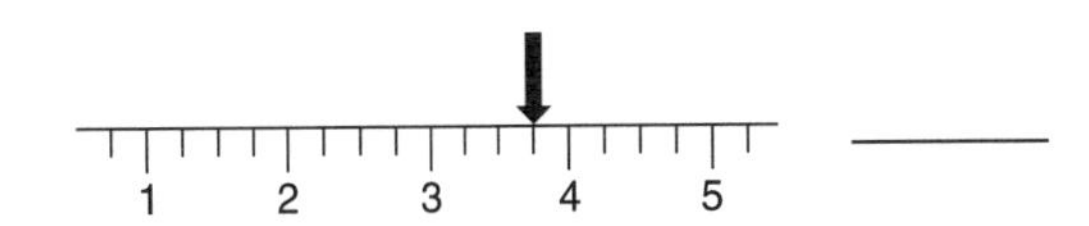

______

The pointer is ________ of the distance from 3 to 4. What number is shown on the scale?

8

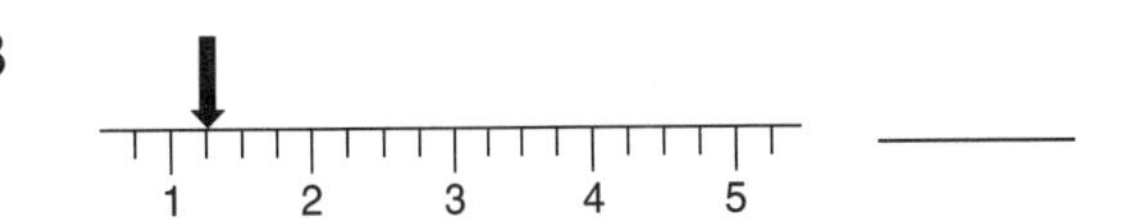

______

What number is shown on the scale?

You can follow these two steps to read a scale.

Step 1 Figure out what each tick mark is worth, as shown at the right.

Step 2 Use the tick marks to count from the lower label to the arrow. For example, if each tick mark is worth two, count by twos; or if each tick mark is worth five, count by fives.

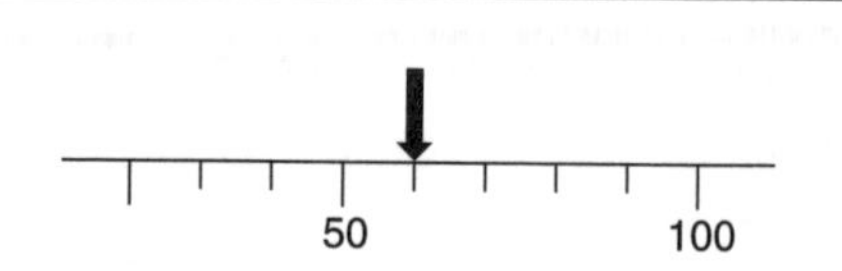

The difference between 50 and 100 is 50. Tick marks divide the interval between 50 and 100 into 5 parts. So each tick mark represents 50 ÷ 5 or 10 units.

The arrow is 1 tick mark more than 50. The arrow points to 50 + 10 or 60.

9

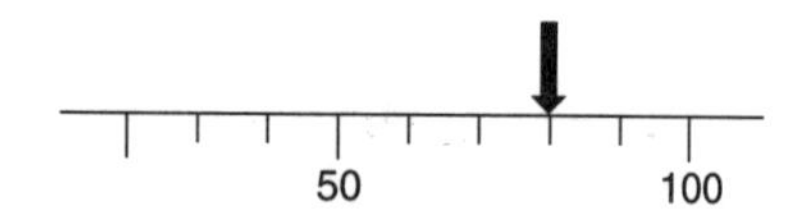

What is each tick mark worth? _______

What number is shown on the scale? _______

10

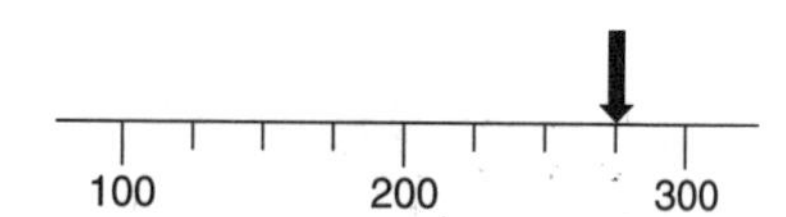

What is each tick mark worth? _______

What number is shown on the scale? _______

11

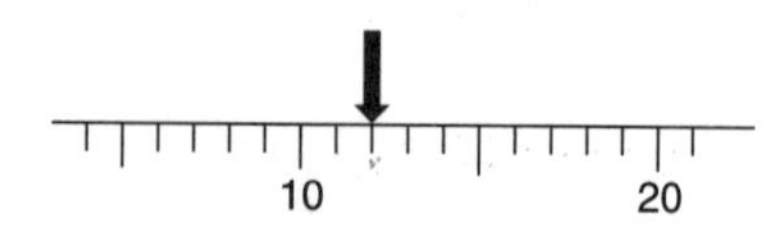

What is each tick mark worth? _______

What number is shown on the scale? _______

12

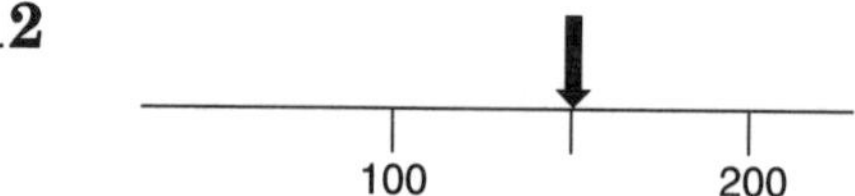

What number is shown on the scale? _______

13

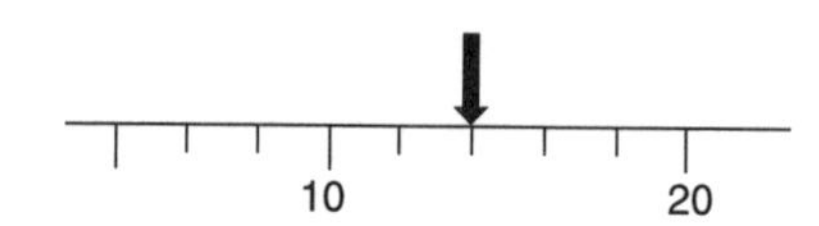

What number is shown on the scale? _______

14

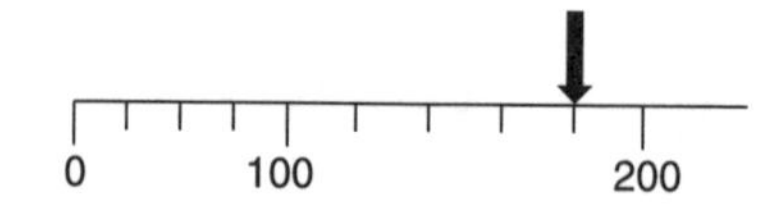

What number is shown on the scale? _______

15

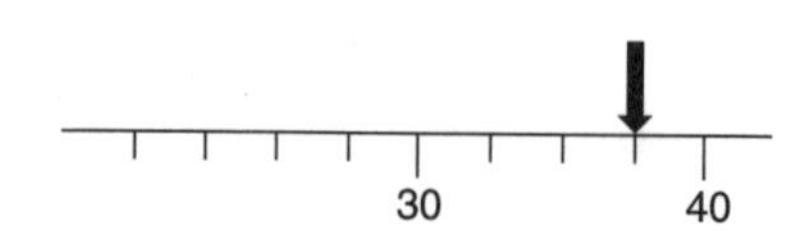

What number is shown on the scale? _______

Estimating a Measurement on a Scale

Sometimes a pointer is between two tick marks. When this happens, you can estimate the reading shown on the scale.

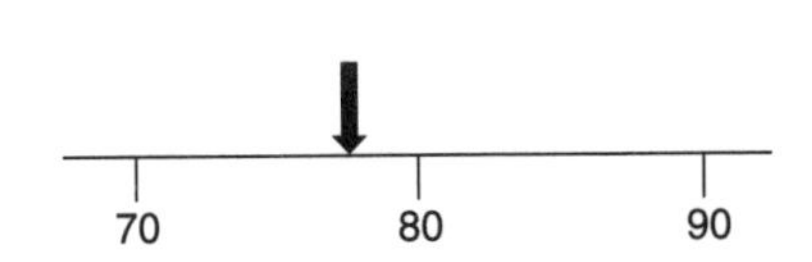

This arrow is between 70 and 80, but it is closer to 80 than to 70.
Estimate: **About 77 or 78.**

There is no one correct answer to this type of an estimation problem. However, 81 would not be a good estimate because 81 is not between 70 and 80. Similarly, 72 would not be a good estimate because 72 is closer to 70 than to 80.

PRACTICE

Circle the letter for the *best* estimate for each measurement.

1

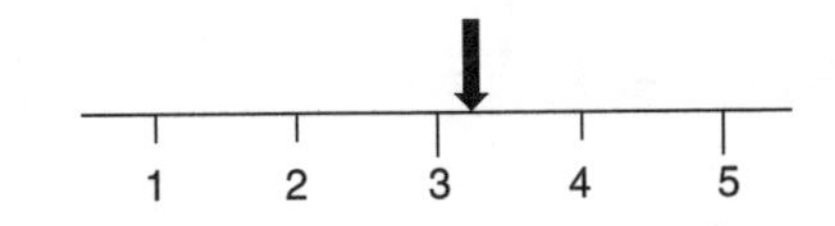

A $3\frac{1}{4}$ **C** $3\frac{3}{4}$

B $3\frac{1}{2}$ **D** 4

2

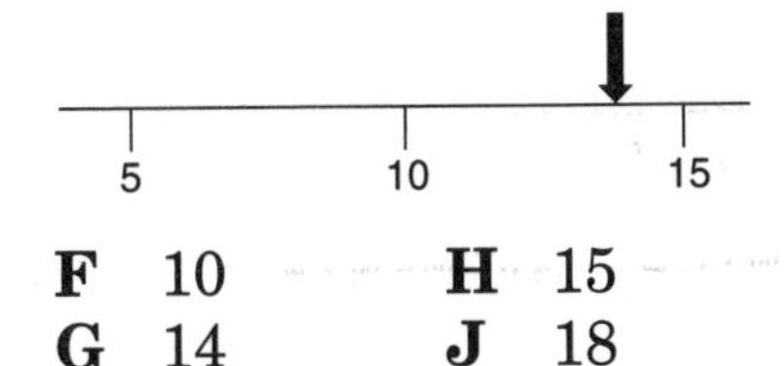

F 10 **H** 15
G 14 **J** 18

3

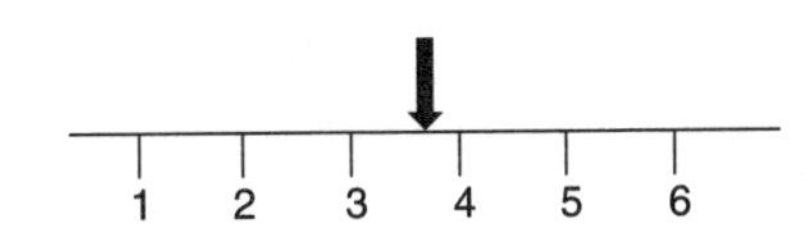

A $2\frac{3}{4}$ **C** 3

B $4\frac{1}{2}$ **D** $3\frac{3}{4}$

4

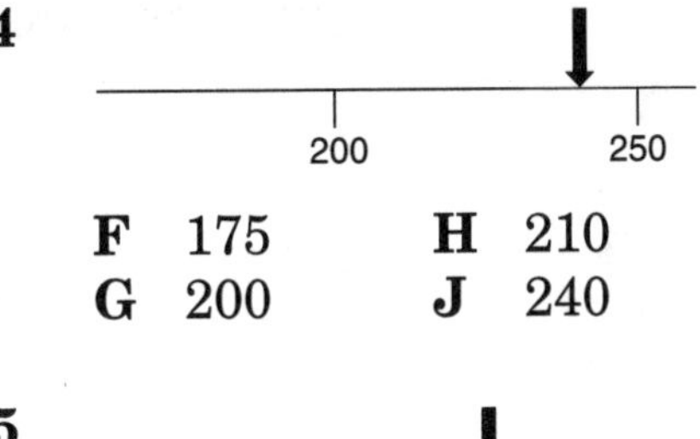

F 175 **H** 210
G 200 **J** 240

5

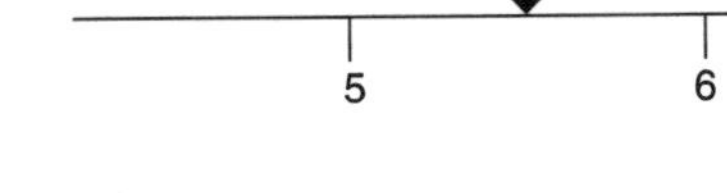

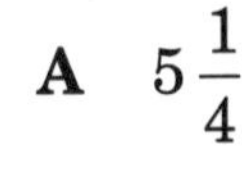

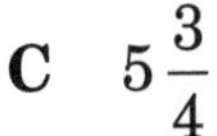

A $5\frac{1}{4}$ **C** $5\frac{3}{4}$

B $5\frac{1}{2}$ **D** 6

6

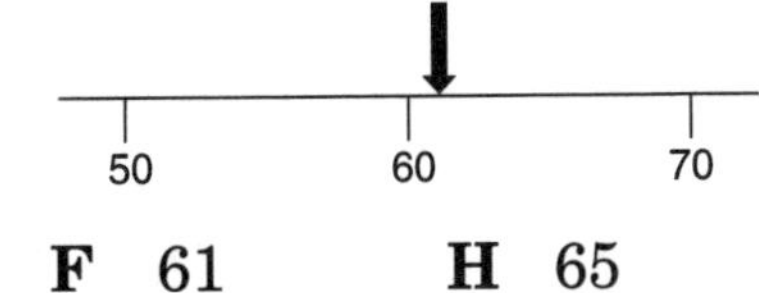

F 61 **H** 65
G 64 **J** 68

Measuring Temperature

Temperature is measured using a **thermometer** marked in degrees *Fahrenheit* (°F) or degrees *Celsius* (°C). Look for one of these two symbols on the thermometer to tell you which units are used.

PRACTICE

Use the thermometer below for questions 1 through 5.

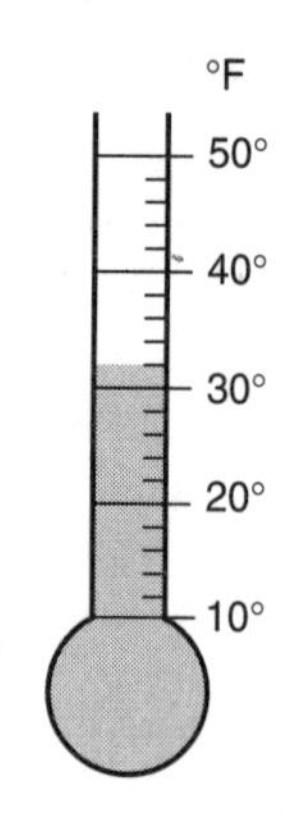

1 Circle one: Which unit does this thermometer use?

degrees Celsius   degrees Fahrenheit

2 What temperature does this thermometer show?

____

3 Is 20°F colder or warmer than the temperature shown above?

____

4 Water freezes at the temperature shown above. Which of these would be the best setting for your freezer?

A 40°F **C** 56°F
B 35°F **D** 5°F

5 Draw a line at 44°F on the thermometer.

The thermometer at the right shows both degrees Celsius and degrees Fahrenheit. Use it for questions 6 through 11.

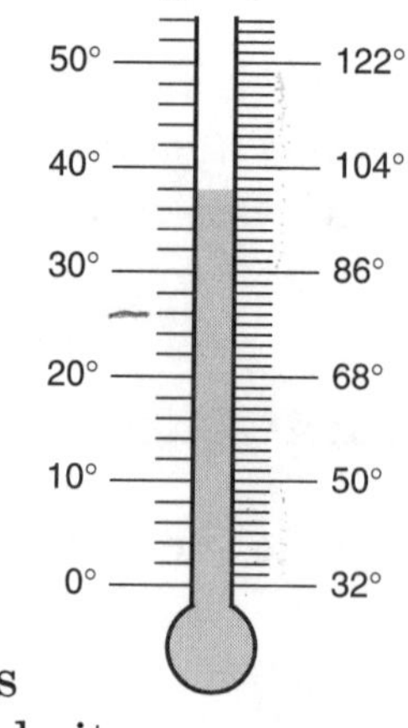

6 Circle one: Which is a larger unit?

one degree Celsius
one degree Fahrenheit

7 Each tick mark on these scales represents how many degrees?

______ °F ______ °C

8 What temperature is shown on this thermometer?

______ °F ______ °C

9 Which of these measurements is the hottest temperature?

A 104°F **C** 40°C
B 50°C **D** 86°F

10 If the outdoor temperature is 42°C, what season is it? (*Hint:* The temperature shown above is normal body temperature.)

____

11 Draw a line at 26°C.

Measuring Length

The distance between two points can be a measurement of width, height, or depth. Each of these measurements is a **length.** The tools below all measure length.

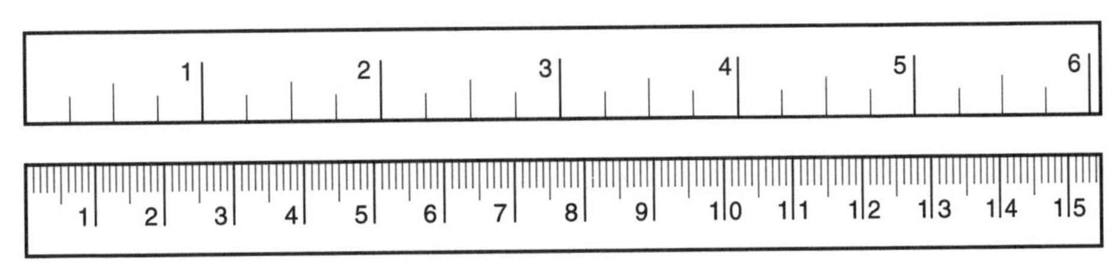

Use a **ruler,** a **yard stick,** or a **meter stick** to measure length on a flat surface.

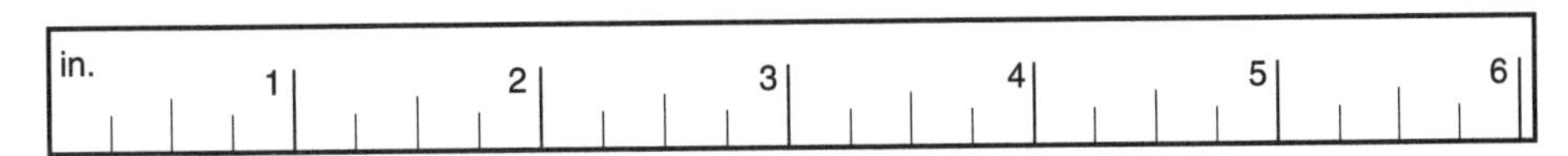

Use a **tape measure** to measure length on curved surfaces or to measure a long distance, like the length of a room.

Standard Units of Length

12 inches (in.)	=	**1 foot (ft)**	An inch is about the length of a straight pin. A foot is about the length of a man's foot.
3 ft	=	**1 yard (yd)**	A yard is about the length of your arm.
5,280 ft	=	**1 mile (mi)**	A mile is about 10 city blocks.

PRACTICE

Circle the letter for the *better* way to measure each length.

1 the height of a wall

A with a ruler
B with a tape measure

2 the length of a pencil

F in inches
G in feet

3 the length of vacant lot

A in miles
B in yards

4 the width of a box

F in inches
G in yards

Use the information below to fill in the blanks.

To change feet to inches, multiply by 12.
To change inches to feet, divide by 12.
To change feet to yards, divide by 3.
To change yards to feet, multiply by 3.

5 A chest is 2 feet tall. How many inches tall is the chest? ________

6 A table is 3 feet tall. How many yards tall is the table? ________

7 How many inches are in $\frac{1}{2}$ a foot? ________

8 A man is 6 feet tall. How tall is the man in yards? ________

9 A foot is what fraction of a yard? ________

The ruler below shows two different scales. One side is marked in inches, which is one of the **standard** units. Other standard units of length are foot, yard, and mile. The other side is marked in centimeters, which is one of the **metric** units. Other metric units of length are millimeters, meters, and kilometers. Abbreviations for the standard units are **in., ft, yd,** and **mi;** abbreviations for the metric units are **mm, m,** and **km.**

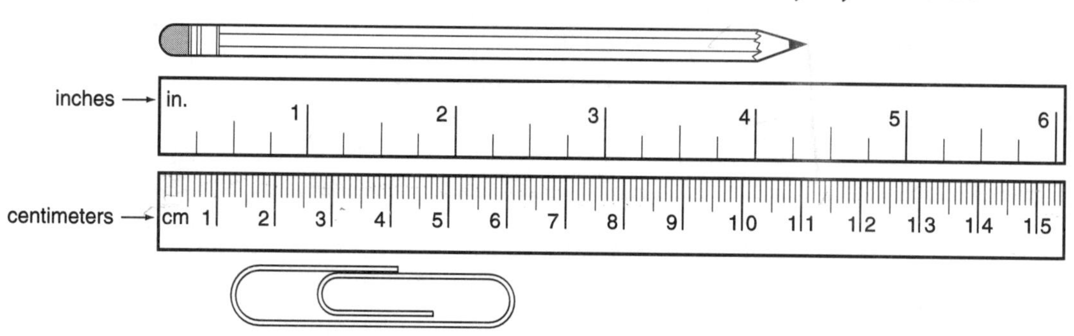

PRACTICE

Use the ruler above to answer each question.

10 About how many centimeters are shown on this ruler? ________

11 About how many inches are shown on the ruler? ________

12 Each centimeter is divided into smaller units, called millimeters. How many millimeters are in each centimeter? ________

13 About how many centimeters are in one inch? *(Estimate to the nearest half centimeter.)* ________

14 In centimeters, about how long is the pencil? ________

15 In inches, about how long is the pencil? *(Estimate to the nearest half inch.)* ________

16 Suppose an object is 3 inches long. About how long is that object in centimeters? *(Estimate to the nearest half centimeter.)* ________

17 Suppose an object is 4 centimeters long. About how long is that object in inches? *(Estimate to the nearest half inch.)* ________

18 How many millimeters are in 7 centimeters? ________

19 On the ruler, draw a line that represents a measurement of 9 centimeters.

20 Look at the paper clip above. Which of these is the best way to measure the paper clip?

A Measure its width rather than its length.
B Turn the ruler over.
C Use the right end of the ruler instead of the left end.
D Line up the left edge of the paper clip with the "zero" mark of the ruler.

Using a Ruler

When you use a ruler, follow these three steps:

1. Line up the zero on the ruler with one end of the object you are measuring.
2. Find the number that lines up with the other end of the object.
3. Measure the object again to check your work.

PRACTICE

Use a ruler marked in inches to measure the length of each bar below. Round each answer to the nearest whole number.

1

_______ inches

2

_______ inches

3

_______ inches

4

_______ inches

5

_______ inches

6

_______ inches

Reading Fractions on Scales

The **tick marks** on a scale divide the units of the scale into smaller parts.

To find out the value of each tick mark, count how many are in a single unit. For example, if a unit is divided into 2 parts, then each part is $\frac{1}{2}$ or 0.5 of a unit. If each unit is divided into 4 parts, then each part is $\frac{1}{4}$ or 0.25 of a unit.

PRACTICE

Measure the length of each bar below to the nearest half-inch. Some of your answers can be whole numbers.

1

_______ inches

2

_______ inches

3

_______ inches

4

_______ inches

Find *any three* objects from the list below. Measure their heights and widths to the nearest half-inch. List the objects you found, and give their measurements.

a check book	a business card
computer paper	a cassette tape
a car key	a news magazine

5 Object: ______________________

The height is _______ inches.

The width is _______ inches.

6 Object: ______________________

The height is _______ inches.

The width is _______ inches.

7 Object: ______________________

The height is _______ inches.

The width is _______ inches.

Measuring to the Nearest $\frac{1}{8}$-inch

For the activities below, find a ruler that divides each inch into at least 8 parts. If you have a ruler that is marked in sixteenths of an inch or even thirty-seconds of an inch, you can use it. Just concentrate on the $\frac{1}{8}$-inch tick marks on the ruler.

PRACTICE

Measure the length of each bar below to the nearest eighth of an inch. You may find the following fraction equations useful.

$\frac{1}{2} = \frac{4}{8}$ $\frac{3}{4} = \frac{6}{8}$

$\frac{1}{4} = \frac{2}{8}$ $\frac{2}{16} = \frac{1}{8}$

1

_______ inches

2

_______ inches

3

_______ inches

4

_______ inches

5

_______ inches

Find *four* of the measurements listed below. Give your answers to the nearest eighth of an inch.

the height or width of a postage stamp: _______ inches

the height or width of a dollar bill: _______ inches

the height or width of a credit card: _______ inches

the height of a business envelope: _______ inches

the width of a large metal paper clip: _______ inches

the length of an AA battery (Ignore the round terminals.): _______ inches

the height of a roll of paper towels: _______ inches

the height of a condensed soup can: _______ inches

Measuring to the Nearest Millimeter

For the activities below, find a ruler that is marked in millimeters. There are about 30 centimeters in a foot, and there are 10 millimeters in each centimeter, so there are about 300 millimeters in a foot.

PRACTICE

Measure the length of each bar below to the nearest millimeter.

1 ▬▬▬▬▬▬▬▬

_______ millimeters

2 ▬▬▬▬▬▬▬▬▬▬▬▬

_______ millimeters

3 ▬▬▬▬▬

_______ millimeters

4 ▬▬▬

_______ millimeters

5 ▬▬▬▬▬▬▬▬▬▬▬

_______ millimeters

Use the same four objects that you measured on the previous page. This time, give your answers to the nearest millimeter.

the height or width of a postage stamp: _______ mm

the height or width of a dollar bill: _______ mm

the height or width of a credit card: _______ mm

the height of a business envelope: _______ mm

the width of a large metal paper clip: _______ mm

the length of an AA battery (Ignore the round terminals.): _______ mm

the height of a roll of paper towels: _______ mm

the height of a condensed soup can: _______ mm

Measuring Weight

For weight, the standard units are **ounces, pounds,** and **tons.** The metric units for weight are **milligrams, grams,** and **kilograms.** Abbreviations for the standard units are **oz, lb,** and **T.** Abbreviations for the metric units are **mg, g,** and **kg.**

1 pound (lb) = 16 ounces (oz) **1 ton (T) = 2,000 lb**	A pencil weighs about 1 ounce. An eggplant weighs about 1 pound. A car weighs about 1 ton.
1 gram (g) = 1,000 milligrams (mg) **1 kilogram (kg) = 1,000 grams (g)**	A needle weighs about 1 milligram. A peanut weighs about 1 gram. A phone book weighs about 1 kilogram.

PRACTICE

Use the table above for questions 1–6.

1 How many ounces are in half of a pound? ______

2 How many grams are in half of a kilogram? ______

3 Which of these would weigh about one ounce?

A a letter **C** a book
B a hamburger **D** a chair

4 Which of these would weigh about one kilogram?

F a penny **H** a pair of shoes
G a couch **J** a sheet of paper

5 What measurement is shown on this scale? ______

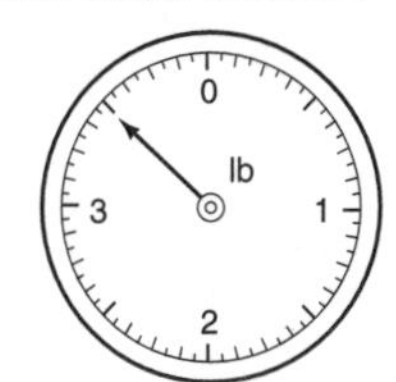

6 What measurement is shown on this scale? ______

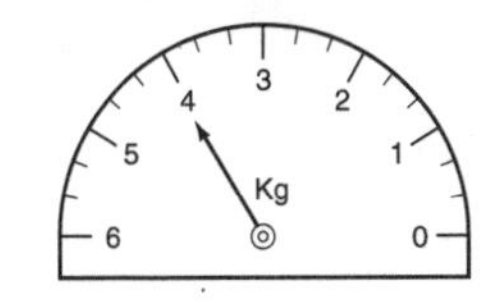

Use the following for questions 7–10.

To change:
ounces into pounds, divide by 16.
pounds into ounces, multiply by 16.
grams to kilograms, divide by 1,000.
kilograms to grams, multiply by 1,000.

7 Suppose you have 3,000 grams of hamburger. How many kilograms of hamburger is that?

______

8 Your kitten weighs 32 ounces. How many pounds is that?

______

9 A package weighs 4 pounds. How many ounces is that?

______

10 You have a 2-kilogram roast. How many grams is that?

______

Circle the heavier weight.

11 9 oz 1 lb

Measuring Liquids

For liquids, the standard units are **cups, pints, quarts,** and **gallons.** The metric units are **liters** and **milliliters.**

1 pint = 2 cups	A small mug holds about 1 cup. A large mug holds about 1 pint.	To change cups into pints, divide by 2. To change pints into cups, multiply by 2.
1 quart = 4 cups (or 2 pints)	A narrow milk carton holds 1 quart.	To change cups into quarts, divide by 4. To change quarts into cups, multiply by 4.
1 gallon = 4 quarts (or 16 cups)	A large milk carton holds 1 gallon.	To change quarts into gallons, divide by 4. To change gallons into quarts, multiply by 4.
1 liter = 1,000 mL	A plastic soda bottle holds 2 liters.	To change milliliters (mL) into liters, divide by 1,000. To change liters into milliliters, multiply by 1,000.

PRACTICE

Use the table at the top of the page for questions 1–6.

1 How many cups are in two pints? ______

2 How many quarts are in two gallons? ______

3 How many milliliters are in 2 liters? ______

4 Tell the number of quarts and the number of gallons in 12 cups.

_______ quarts _______ gallons

5 About how much liquid is in a can of soda?

A 1 cup **C** 10 milliliters
B 1 quart **D** 1 liter

6 A pitcher holds 1–2 units of water. What is the most likely unit?

F pint **H** gallon
G quart **J** milliliter

Use this diagram for questions 7–11.

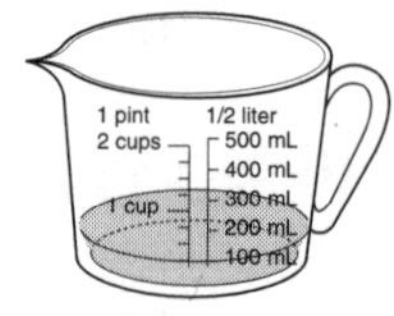

7 How many pints are shown on this scale? _______

8 How much liquid is shown?

_______ cup about _______ mL

9 Suppose a glass contains 500 milliliters of milk. How many liters of milk is that? _______

10 About how many milliliters are in one cup? _______

11

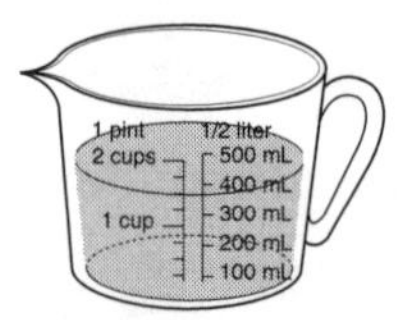

How many cups of liquid are shown here? _______

Reading and Writing Time

The clock at the right shows the time: forty-three minutes after four o'clock. Using figures, that time is written "4:43," just as it would appear on a digital clock. The number to the left of the colon is the hour. The number to the right of the colon represents minutes. When you say a time aloud, you say the hour and then the minutes.

4:43
four forty-three

On a clock face, the numbers 1 through 12 name the hours, not the minutes. One way to find the number of minutes is to remember that each number stands for 5 minutes. If the minute hand (the longer hand) points to 1, it is 1×5 or 5 minutes after the hour. For example, when the minute hand points to 2 it is 2×5 or 10 minutes after the hour.

PRACTICE

Write each time in words and numbers.

Sample.

Numbers: <u>5:10</u>
Words: <u>Five ten</u>
(or ten minutes after five)

1

Numbers: _______

Words: ____________________

2

Numbers: _______

Words: ____________________

3

Numbers: _______

Words: ____________________

Use the information below for questions 4–8.

1 minute	**=**	**60 seconds**
1 hour	**=**	**60 minutes**
24 hours	**=**	**1 day**

4 How many minutes are in one-half of an hour? _______

5 How many minutes are in one-quarter of an hour? _______

6 $1\frac{1}{2}$ hours is _______ minutes, or 1 hour and _______ minutes.

7 How many hours are in 120 minutes? _______

8 How many hours are in one half of a day? _______

Changes in Time

Here are two problems that deal with starting and finishing times. To solve them, imagine how the hands of a clock would move. First, figure out the change in hours. Then figure out the change in minutes:

Problem 1

A movie is 2 hours and 10 minutes long. If you start watching it at 7:30, when will it be over?

You need a time *after* 7:30, so count forward.
Count the hours:
Two hours after 7:30 is 9:30.
Count the minutes:
10 minutes after 9:30 is 9:40.

Problem 2

Shu has a 10:00 doctor's appointment. It will take 1 hour and 15 minutes to get to the doctor's office. Shu must leave by what time?

You need a time *before* 10:00, so count backward.
Count the hours:
1 hour before 10:00 is 9:00.
Count the minutes:
15 minutes before 9:00 is 8:45.

PRACTICE

Fill in the blank for each problem.

1 What time is 2 hours before 1:00?

______

2 Suppose that potatoes will take one and a half hours to bake. You plan to eat at 5:00. By what time should you should start the potatoes?

______

3 It will take one hour and forty-five minutes for an airplane flight. You take off at 4:30. At what time do you land?

______

4 What time is three and a half hours after 10:00?

______

5 It should take 5 hours and 15 minutes for the turkey to roast. If you start it at 8:00, when will it be done?

______

6 You just finished a 45-minute workout. It is now 2:15. When did you start?

______

This problem asks you to find the **duration** between a starting time and the finishing time. First, count the number of full hours that have passed. Then, count the number of minutes that have passed.

Eric started work at 8:25. He finished at 4:10. How long did he work?

Starting time

Ending time

Count the number of hours. He did not work as late as 4:25, so 7 full hours passed. (That takes him to 3:25.)
Count the number of minutes. From 3:25 to 4:00 is 35 minutes, and from 4:00 to 4:10 is 10 minutes.
Answer: 7 hours and 45 minutes.

PRACTICE

Find the duration between each starting time and ending time.
(Each duration is *less than* 12 hours.)

7

Starting Time

Ending Time

Time passed ________________

8

Starting Time

Ending Time

Time passed ________________

9 Randi starts driving at 2:10. She arrives at 8:30. How long does her trip take?

____

10

Starting Time

Ending Time

Time passed ________________

11 At 11:00, you drop off your children at the mall. At 1:30, they call to say they are ready to come home. How much time has passed?

____

12 A track meet starts at 9:30. It ends at 6:00. How long does it last?

____

Reading a Map or Diagram

An inch on a map or diagram usually stands for a particular distance. Look for a label or key that tells you about the actual distances.

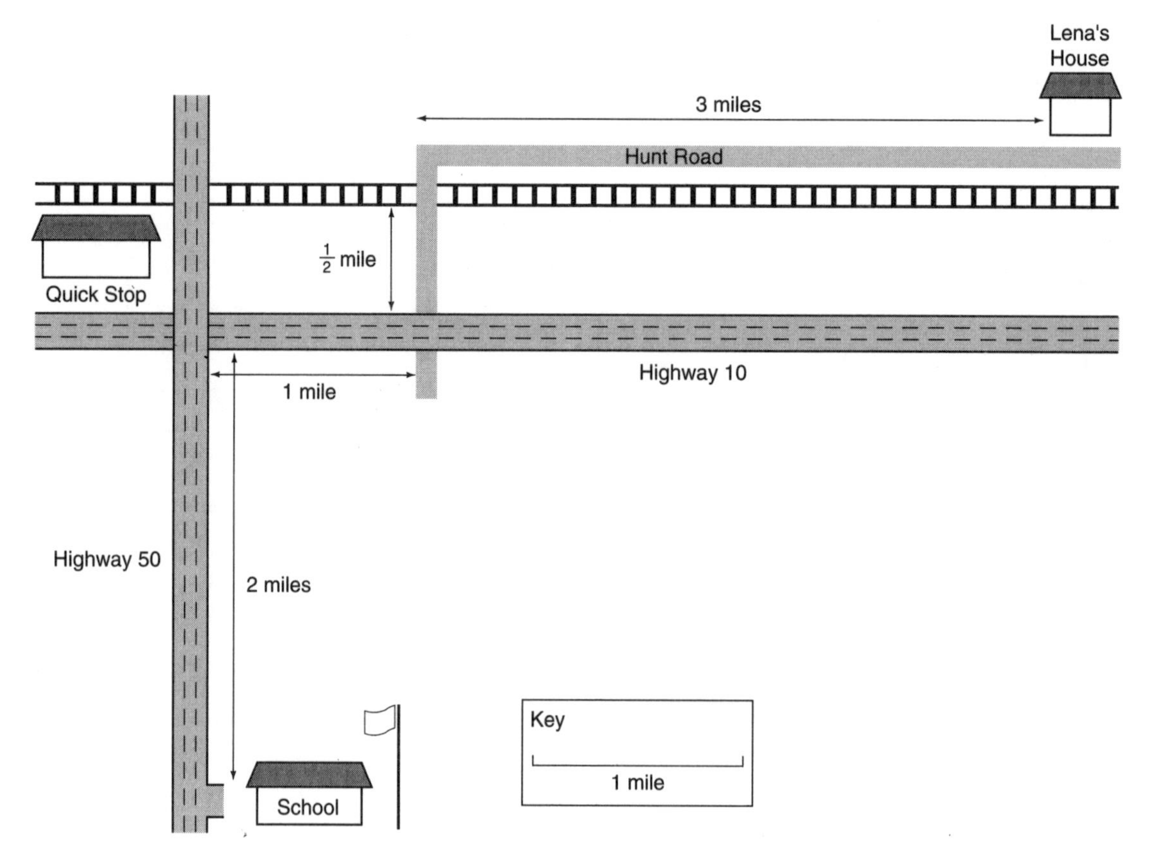

PRACTICE

Use the map above to answer these questions.

1 On this map, an inch represents what distance?

A 1 yard　**C** 1 inch
B 1 foot　**D** 1 mile

2 Lena works at the Quick Stop. How far does she have to drive to work?

____

3 Lena's children meet her at the Quick Stop after school. How far do they walk?

____

4 Which drive is farther?

A Lena's house to the school.
B Lena's house to the Quick Stop

5 It takes Lena's children 30 minutes to walk to the Quick Stop from school. If school is out at 3:30, at what time do they reach the Quick Stop?

____

6 Lena leaves for work at 8:10. She arrives at 8:22. How long does it take Lena to get to work?

____

Measurement Skills Practice

Circle the letter for the correct answer to each problem.

1 How much liquid is shown?

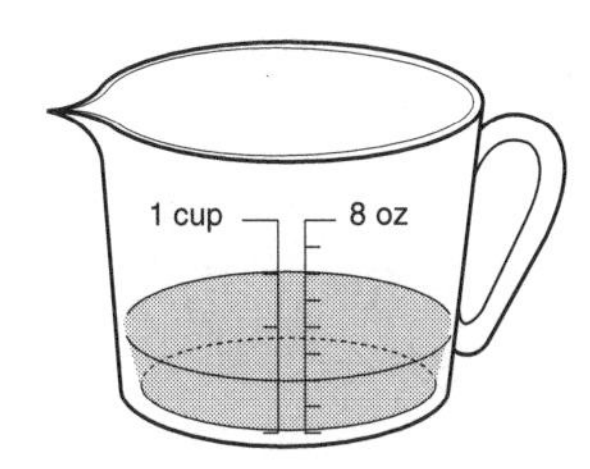

A $\frac{1}{4}$ cup   **C** 2 cups
B 1 cup   **D** 3 cups

2 How many feet are in 2 yards?

F 3   **H** 20
G 6   **J** 24

3 What time is 4 hours before 4:00?

A 1:00   **C** 11:00
B 2:00   **D** 12:00

4 What temperature is shown on this thermometer?

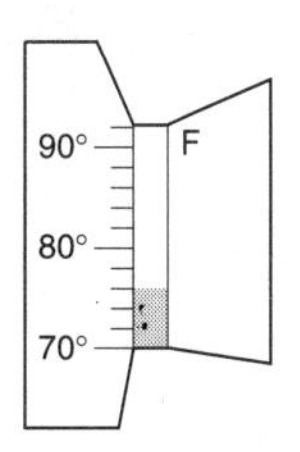

F 72 °F   **H** 75 °F
G 73 °F   **J** 76 °F

5 How many minutes are in one quarter of an hour?

A 10   **C** 20
B 15   **D** 45

6 Which would be the best tool to use to measure the length of a truck?

F

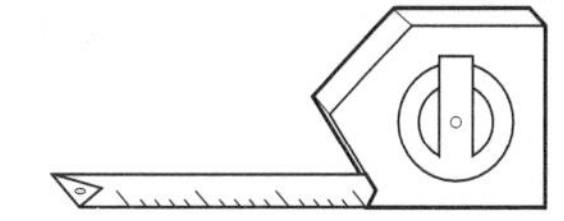

G   **H**

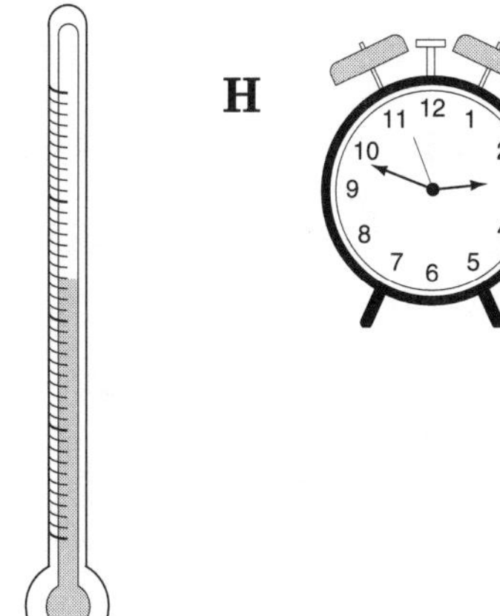

J

7 Lynn has a shelf that is 15 inches wide. She fills 9 inches of the shelf with books. How much space is left for other things?

A 5 inches
B 4 inches
C 6 inches
D 1 foot

8 Which of these tools could you use to measure 3 pounds of sand?

F a measuring cup
G a meter stick
H measuring spoons
J a bathroom scale

9 Which group of measurements is in order from shortest to longest?

A 5 inches, 1 foot, $\frac{1}{2}$ foot

B 1 foot, $\frac{1}{2}$ foot, 5 inches

C $\frac{1}{2}$ foot, 5 inches, 1 foot

D 5 inches, $\frac{1}{2}$ foot, 1 foot

10 The train trip from Detroit to Kalamazoo takes $2\frac{1}{2}$ hours. If the train leaves Detroit at 2:00, what time will it arrive in Kalamazoo?

F 3:30
G 5:30
H 4:30
J 6:30

11 The photographer says Ellyn's pictures will be ready in 21 days. How many weeks is that?

A 2
B 3
C 4
D $2\frac{1}{2}$

12 Akmed is 6 feet tall. He buys a top hat that extends 6 inches above the top of his head. How tall is he when he wears the hat?

F 7 feet tall

G $6\frac{1}{2}$ feet tall

H $6\frac{1}{4}$ feet tall

J $7\frac{1}{2}$ feet tall

13 A truck driver drove 605 miles on Monday, 490 miles on Tuesday, and 503 miles on Wednesday. How many miles did he drive in all?

A 1,598
B 1,517
C 1,498
D 1,599

14 Yesterday, Qiu worked from 6:00 in the morning until 5:30 at night. How many hours did he work?

F $1\frac{1}{2}$

G $5\frac{1}{2}$

H $9\frac{1}{2}$

J $11\frac{1}{2}$

Geometry

Using Logic

Solve these problems step-by-step. Make sure your thinking is correct before you move on to the next step. Make notes and pictures to organize your thoughts.

PRACTICE

Solve each problem below. *Hint:* First, cross out all the answers that *could not* be correct.

1 Which number is in all three circles *and* is less than 5?

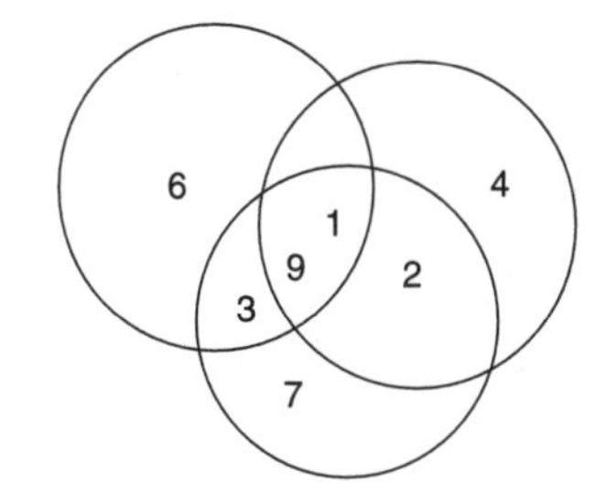

____

2 Circle the figure that has an even number of sides *and* is smaller than the dark circle.

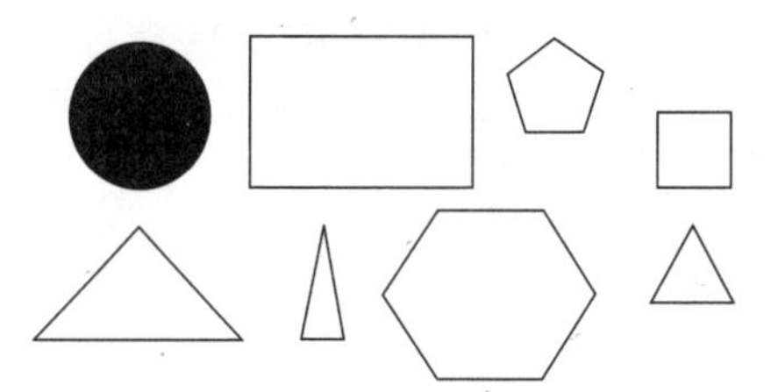

3 Which letter is in the circle and the square, but *not* in the triangle?

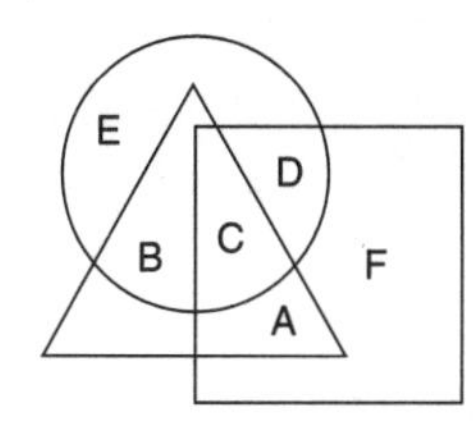

____

4 Which shape contains the star and has four sides?

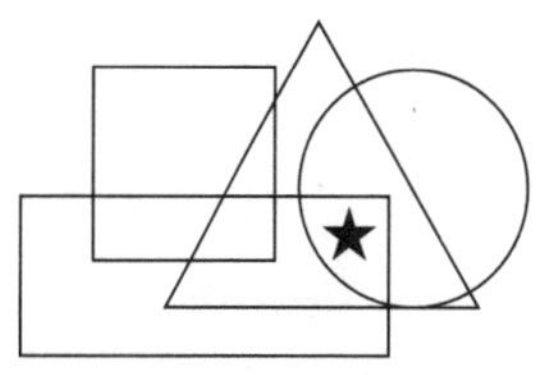

F the circle
G the square
H the triangle
J the rectangle

5 Which square has dots and stripes, but *no* stars?

A **B** **C** **D**

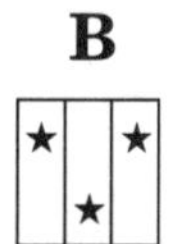
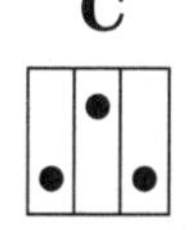
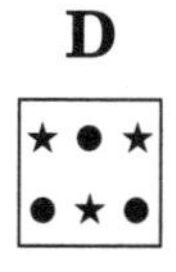

6 Which of these rectangles is twice as long as it is tall?

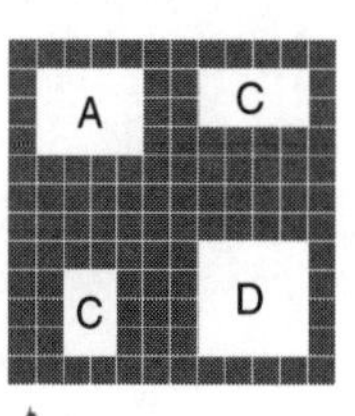

____

Figures

In geometry you investigate lines, angles, and shapes. This table shows four common shapes.

Shape	Examples	Meaning
circle		A figure with all points a *particular distance* from a *center* point
quadrilateral		a four-sided figure
square		a figure with 4 equal sides and 4 equal angles
triangle		a three-sided figure

PRACTICE

Each problem below contains facts about a shape, followed by a conclusion in bold type. The facts are always correct. The conclusion is not always correct. On each blank line, tell whether the conclusion is "correct" or "incorrect."

1

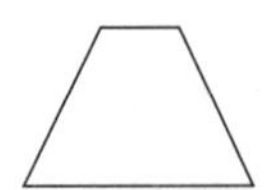

A triangle is a figure with three sides. This figure has four sides. **Therefore, this is not a triangle.**

____

2 A quadrilateral is a figure with 4 sides. A square is a figure with 4 equal sides and 4 equal angles. **Therefore, all squares are quadrilaterals.**

____

3 A square is a 4-sided figure whose sides are all the same length. **Therefore, any figure that has four sides must be a square.**

____

4

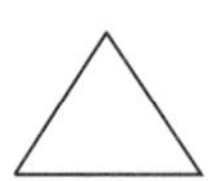

All three sides of this triangle are the same length. One side is 2 inches long. **Therefore, the distance around the triangle must be 2×3 or 6 inches.**

____

5

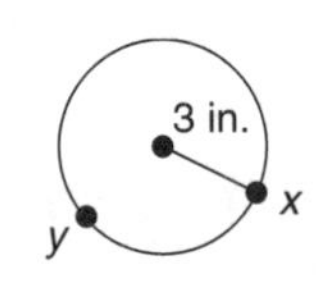

Every part of this circle is 3 inches from the center. **Therefore, point *x* must be 3 inches from point *y*.**

____

Drawing Figures

An important skill in geometry is to draw or be able to imagine shapes. Also, you need to be able to look at shapes from different points of view.

PRACTICE

For problems 1 through 9, draw or imagine each figure. Use your drawn or imagined figure to solve the problem.

1 If you were to draw a line from corner *A* to corner *B*, it would create two __?__.

B

A

A quadrilaterals
B triangles
C squares
D none of these

2 A piece of paper has been folded in half and then cut as shown. Once it is unfolded, what is the shape of the cut-out?

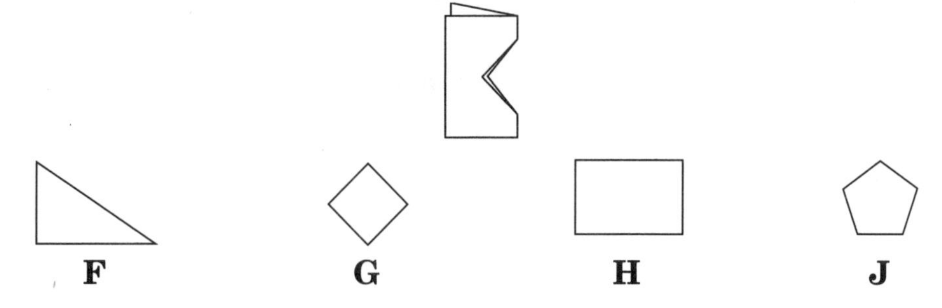

3 This figure is like the center of a roll of paper towels. It is called a **cylinder.** If you trace the bottom of the cylinder, what shape do you draw?

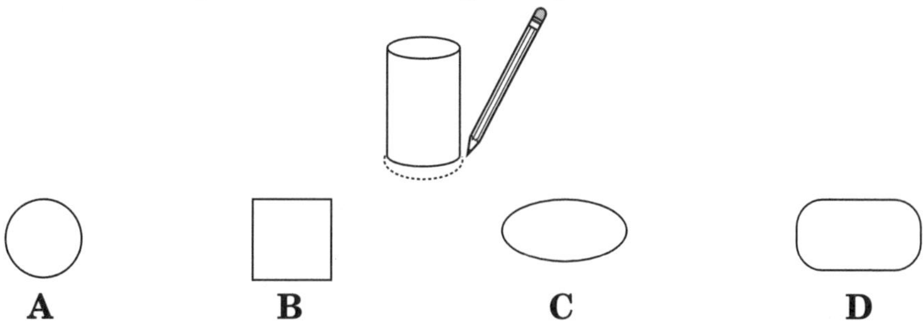

4 What is the shape of this cut-out?

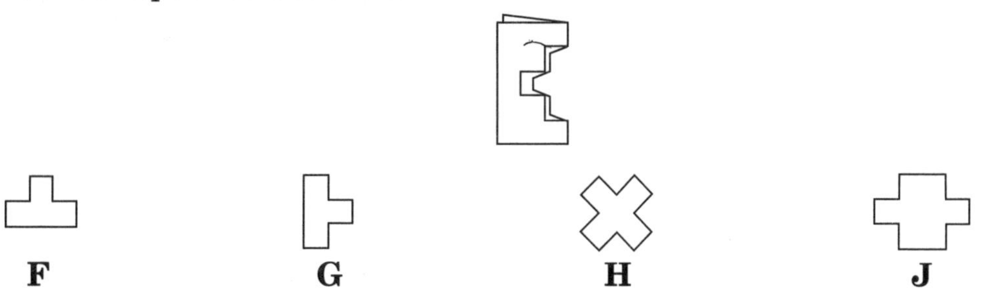

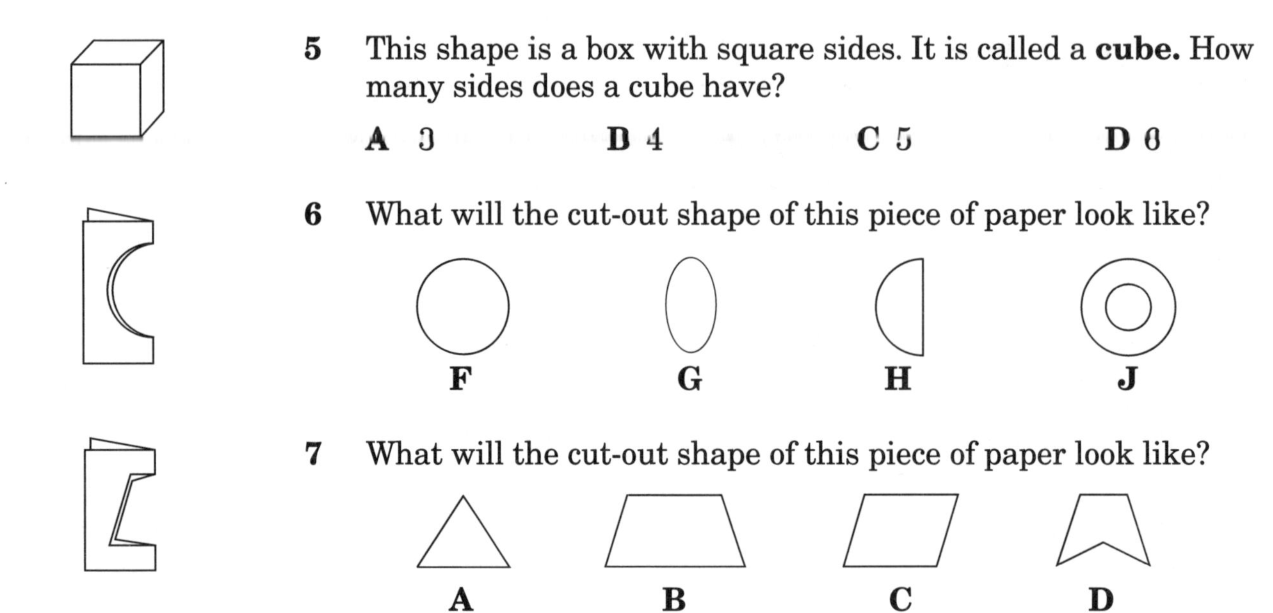

5 This shape is a box with square sides. It is called a **cube.** How many sides does a cube have?

A 3 **B** 4 **C** 5 **D** 6

6 What will the cut-out shape of this piece of paper look like?

F **G** **H** **J**

7 What will the cut-out shape of this piece of paper look like?

A **B** **C** **D**

8 This figure is the shape of a ball, called a **sphere.** When a sphere is resting on a table, what does the part of the sphere that touches the table look like?

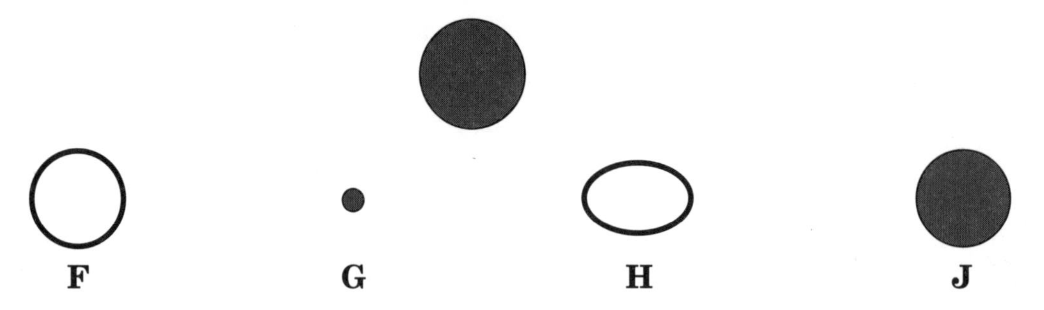

F **G** **H** **J**

9 This shape has a bottom, and it has three sides that are triangles. The shape is called a **pyramid.** If you trace the bottom of this pyramid, what shape would you draw?

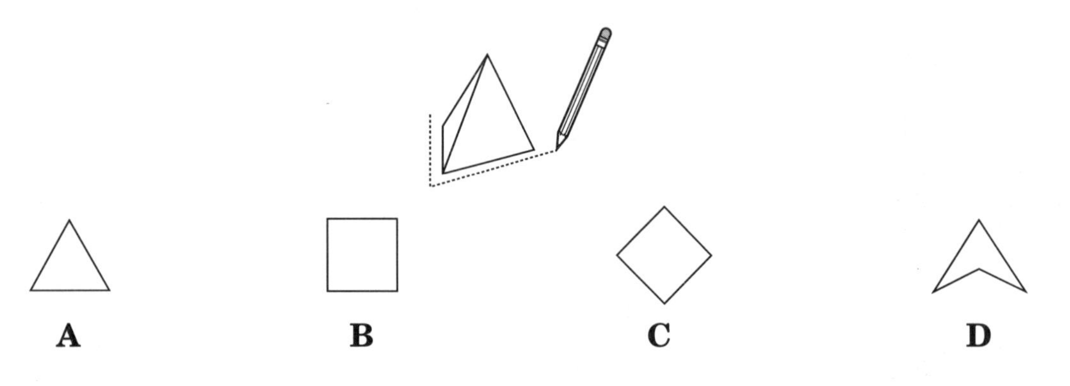

A **B** **C** **D**

Figures with the Same Shape

PRACTICE

For each problem, look at the figure in the dark box. Then circle the letter for the other figure that has the same *size* and the same *shape*. (Don't be fooled. All the figures have been turned.)

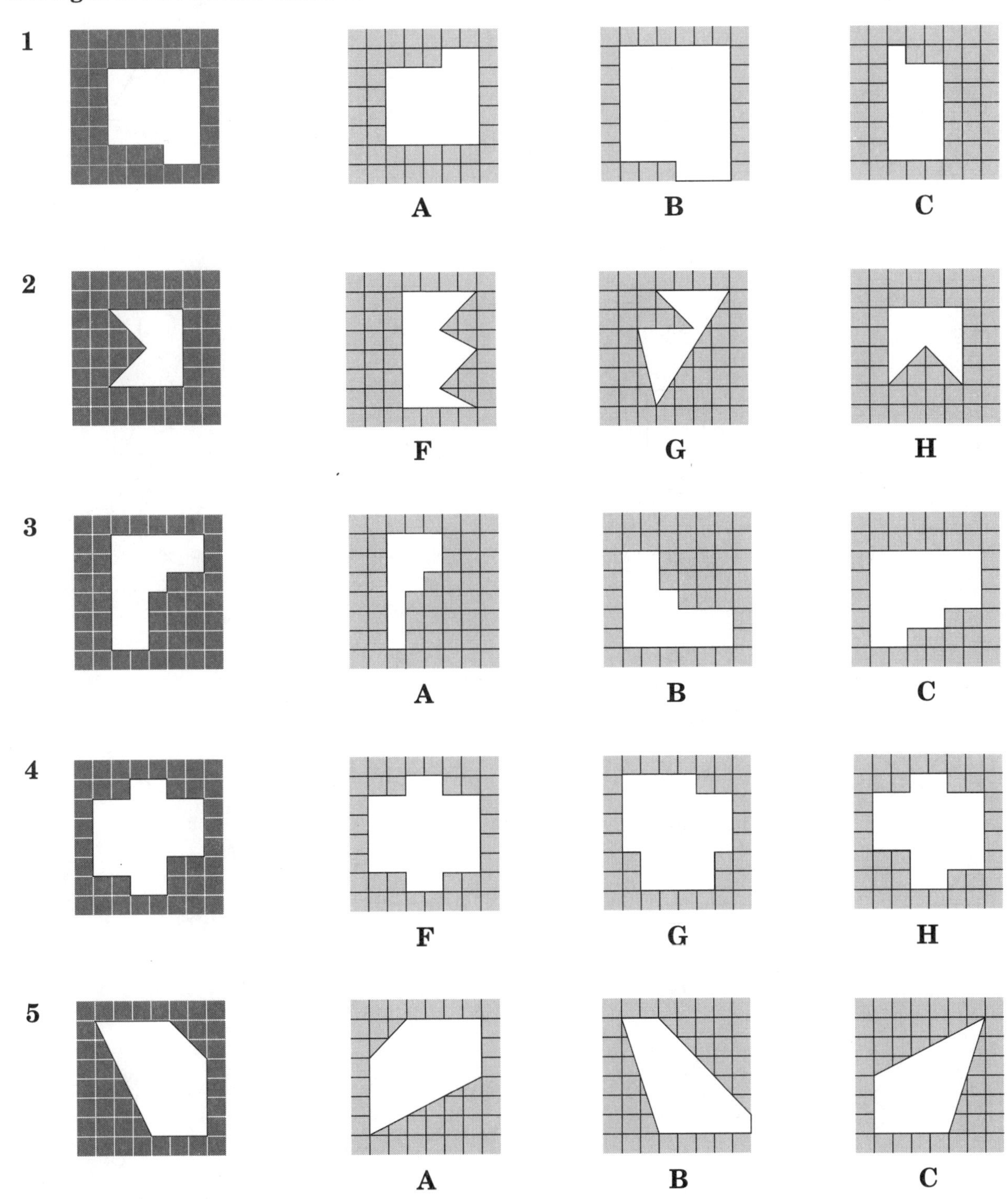

Figures and Symmetry

PRACTICE

Each problem has four copies of the same figure. In each copy, a dashed line forms two smaller figures. Circle the letter where the two smaller figures have the same shape and same size.

1

A B C D

2

F G H J

3

A B C D

4

F G H J

5

A B C D

6

F G H J

7

A B C D

Geometry Skills Practice

Circle the letter for the correct answer to each problem.

1 Which number is even and is outside the oval?

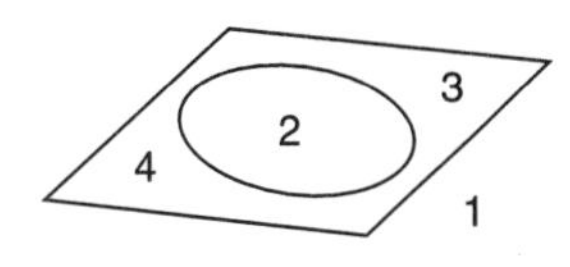

A 1 **B** 2 **C** 3 **D** 4

2 Which number is less than 3 and is outside the figure?

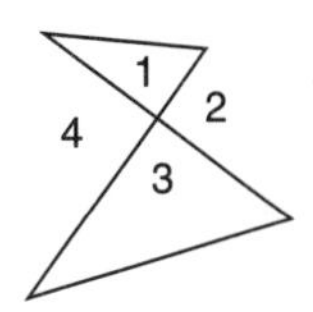

F 1 **G** 2 **H** 3 **J** 4

3 Which figure has the same size and same shape as the figure in the box?

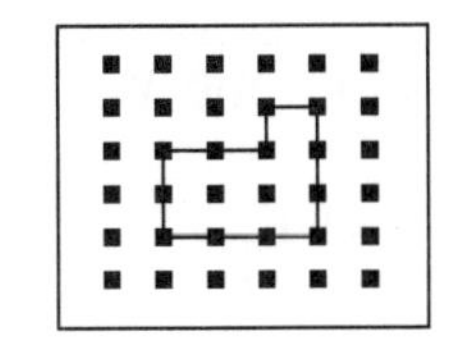

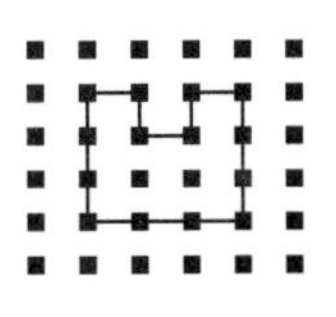

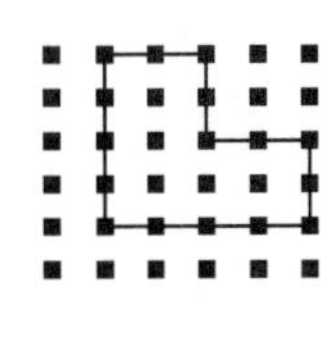

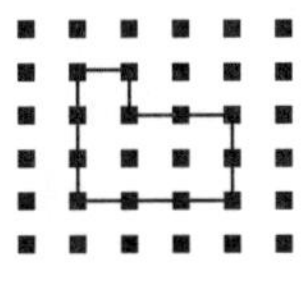

A **B** **C** **D**

4 Which of these figures can be cut along the dashed line so that the two smaller figures match?

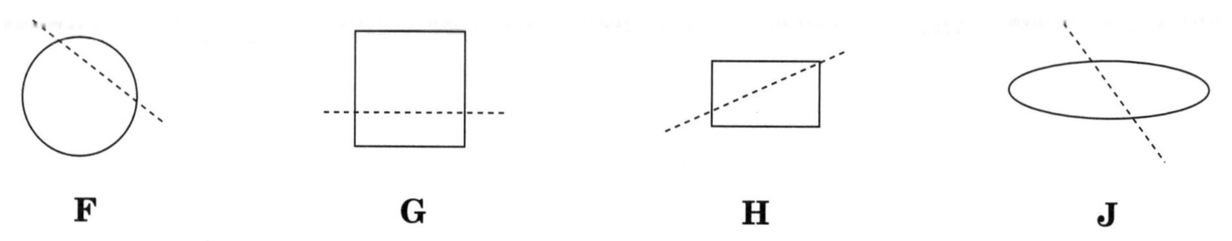

5 Which figure do you draw if you trace the bottom of the cube?

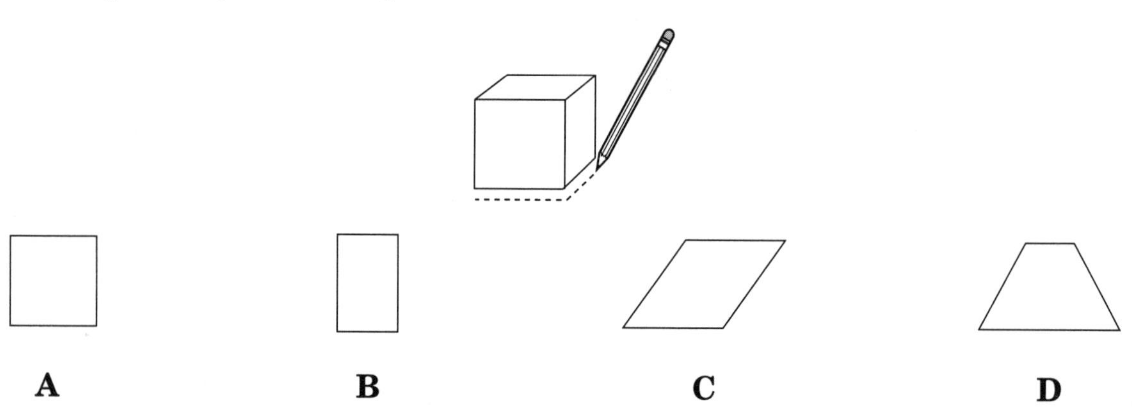

6 A piece was cut out of the folded paper. Which of these shows the cut-out?

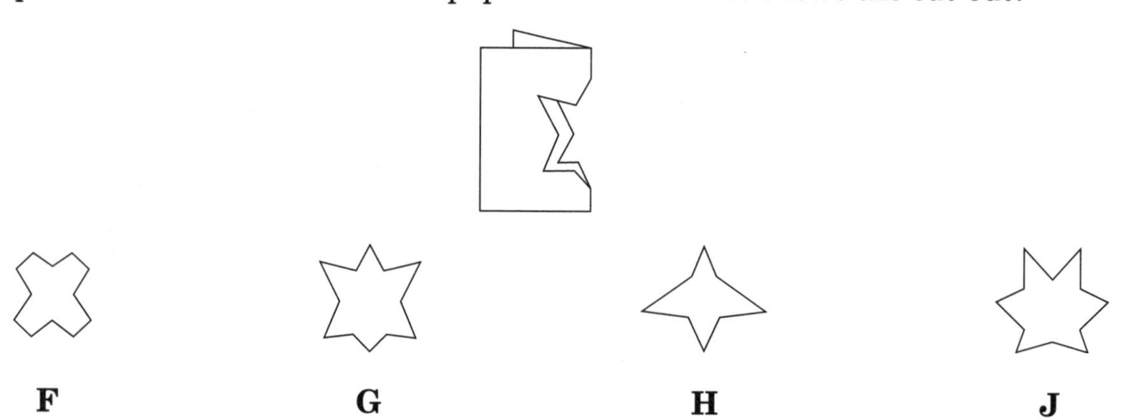

Skills Inventory Post-Test

Part A: Computation

Circle the letter for the correct answer to each problem.

1

$$\begin{array}{r} 35 \text{ in.} \\ + \ 24 \text{ in.} \\ \hline \end{array}$$

A 59 in.
B 11 in.
C 51 in.
D 49 in.
E None of these

2

$$\begin{array}{r} 21 \\ 23 \\ 40 \\ + \ 4 \\ \hline \end{array}$$

F 21
G 67
H 103
J 63
K None of these

3

$90 + 90 =$ ____

A 180
B 160
C 1,800
D 1,600
E None of these

4

$$\begin{array}{r} 282 \\ - \ 180 \\ \hline \end{array}$$

F 102
G 182
H 100
J 82
K None of these

5

$38 - 35 =$ ____

A 2
B 4
C 13
D 30
E None of these

6

$56 - 44 =$ ____

F 21
G 10
H 11
J 12
K None of these

7

$$\begin{array}{r} 10 \\ \times \ 5 \\ \hline \end{array}$$

A 15
B 50
C 55
D 5
E None of these

8

$7 \times 8 =$ ____

F 37
G 47
H 42
J 56
K None of these

9

$$\begin{array}{r} 9 \\ \times \ 4 \\ \hline \end{array}$$

A 13
B 32
C 35
D 45
E None of these

10

$101 + 79 =$ ____

F 180
G 169
H 178
J 170
K None of these

11

$$\begin{array}{r} 939 \\ + \ 109 \\ \hline \end{array}$$

A 148
B 1,048
C 138
D 1,038
E None of these

12

$$\begin{array}{r} 243 \\ 210 \\ + \ 103 \\ \hline \end{array}$$

F 546
G 456
H 556
J 553
K None of these

13

$$189 + 167$$

A 357
B 346
C 256
D 356
E None of these

14

$3\overline{)21}$

F 9
G 7
H 8
J 6
K None of these

15

$9\overline{)72}$

A 6
B 7
C 9
D 8
E None of these

16

$$762 - 660$$

F 102
G 162
H 120
J 112
K None of these

17

$$331 - 304$$

A 27
B 33
C 97
D 37
E None of these

18

$$387 - 199$$

F 88
G 188
H 89
J 198
K None of these

19

$802 - 9 =$ ____

A 803
B 703
C 893
D 793
E None of these

20

$$23 \times 2$$

F 26
G 45
H 46
J 25
K None of these

21

$$18 \times 8$$

A 84
B 144
C 94
D 864
E None of these

22

$$109 \times 5$$

F 514
G 145
H 95
J 545
K None of these

23

$4\overline{)640}$

A 110
B 16
C 160
D 110 r 2
E None of these

24

$300 \div 7 =$ ____

F 43
G 42
H 42 r 4
J 42 r 6
K None of these

25

$5\overline{)210}$

A 42 r 3
B 420
C 240
D 4,200
E None of these

Part B: Applied Mathematics

Circle the letter for the correct answer to each problem.

1 Which problem below has the same answer as the problem in the box?

6 − 0 − 2 =

A 6 − 2 =
B 2 − 0 − 6 =
C 6 + 2 =
D 2 + 6 =

2 What temperature does this thermometer show?

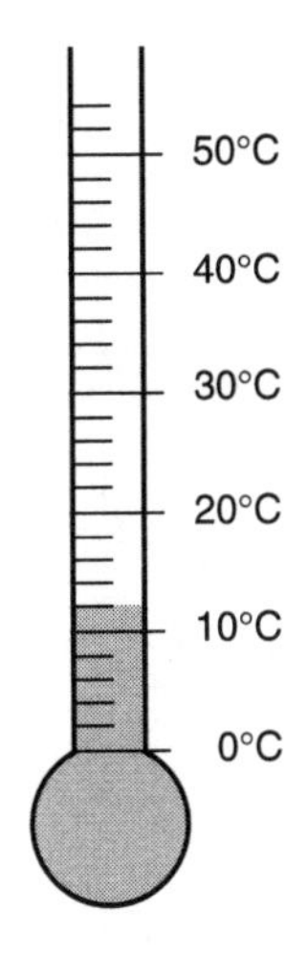

F 10°C
G 11°C
H 12°C
J 15°C

3 What number goes in the boxes to make both number sentences true?

5 + ☐ = 13
13 − ☐ = 5

A 9
B 3
C 6
D 8

4 What number is odd and is inside the circle?

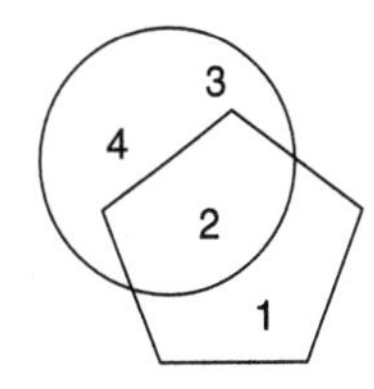

F 1
G 2
H 3
J 4

5 Which of the numbers below is another name for eight hundreds two ones?

8 hundreds 2 ones

A 82
B 820
C 802
D 8,002

6. What number comes before 12 when you count by threes?

F 10
G 9
H 8
J 11

7 Which group of numbers is in order from smallest to largest?

A 11, 32, 25, 18, 29
B 11, 18, 29, 25, 32
C 11, 18, 25, 29, 32
D 18, 11, 25, 29, 32

This table shows how many trucks, minivans, and cars a salesperson sold in May and June. Study the table. Then do Numbers 8 through 11.

Joseph's Sales in May and June

	Trucks	**Minivans**	**Cars**
Sales in May	**3**	**8**	**25**
Sales in June	**6**	**9**	**22**

8 How many cars did Joseph sell in June?

F 9
G 6
H 25
J 22

9 How many trucks did Joseph sell altogether in May and June?

A 9
B 6
C 3
D 33

10 Joseph's May truck sales were what fraction of his June truck sales?

F $\frac{3}{6}$ **H** $\frac{1}{3}$
G $\frac{2}{3}$ **J** $\frac{6}{3}$

11 Which type of vehicle did Joseph sell most often?

A trucks
B minivans
C cars
D There is no way to tell.

Maya has a new job renting paddle boats at the lake. The chart on the left shows how many boats she rented each hour on her first day. The table on the right shows the rent she charged. Study the chart and the table. Then do numbers 12 through 17.

Hour	Boats Rented
1st	✓ ✓ ✓ ✓ ✓ ✓
2nd	✓ ✓ ✓ ✓
3rd	✓ ✓
4th	✓ ✓ ✓ ✓ ✓ ✓ ✓ ✓
5th	✓ ✓ ✓

12 In what hour did Maya rent the most boats?

F the 1st
G the 2nd
H the 3rd
J the 4th

13 How many boats did Maya rent in her 2nd hour on the job?

A 6
B 4
C 2
D 8

14 This equation shows how much Maya earns for 5 hours of work:

$5 \times W = 30$ dollars

W is Maya's hourly wage. What is her hourly wage?

F 7 dollars
G 6 dollars
H 5 dollars
J 8 dollars

Charges for Boat Rental

Boat Size	Time		
	½ hr	1 hr	1 ½ hr
single	$1.50	$2.00	$2.50
double	$2.50	$3.00	$3.50

15 Somebody wants to rent a double boat for 2 hours. If Maya sticks to the pattern shown on this table, how much should she charge?

A $3.50
B $3.00
C $6.00
D $4.00

16. One group rents 2 single boats and 1 double boat for $\frac{1}{2}$ hour each. What should Maya charge?

F $4.00
G $3.00
H $5.00
J $5.50

17 A group rents a single boat for $1\frac{1}{2}$ hours. They pay with a 5-dollar bill. How much change should Maya give them?

A $2.00
B $2.50
C $3.50
D $1.50

Use this advertisement to do Numbers 18 through 23.

Starting Thursday morning at 7:00
Rocky's Annual 13-hour sale
Everything marked down

Bath Rugs	**$9.95**
Handbags	**$12.75**
Watches	**$15.25**
Tights (2 pair)	**$8.40**
Socks (1 pair)	**$4.10**
Twill Pants	**$19.95**
Jeans	**$28.50**

18 What is the sale price on a pair of jeans?

F $9.95
G $19.95
H $28.50
J It is impossible to tell.

19 About how much would it cost to buy two pair of twill pants on sale?

A $20.00
B $30.00
C $40.00
D $50.00

20 How much would 1 pair of tights cost?

F $5.40
G $4.20
H $2.20
J $6.40

21 About how much would it cost to buy a pair of socks and a handbag?

A $17.00
B $18.00
C $16.00
D $20.00

22 When will the sale end?

F 8:00 Friday morning
G 6:00 Thursday evening
H 10:00 Thursday evening
J 8:00 Thursday evening

23 During this sale, the store takes in three thousand, six hundred forty-two dollars. Which of these numbers shows three thousand, six hundred forty-two?

A 30,642
B 36,042
C 3,642
D 306,042

Read this passage. Then do Numbers 24 through 28.

This year the girls club raised money by growing popcorn. They grew and bottled 86 jars.

24 A cardboard box holds 8 jars of popcorn. How many boxes does the club need to hold all the jars they have?

F 8
G 9
H 10
J 11

25 The movie theater has ordered 65 jars. About how many jars does the club have left to sell?

A 20
B 25
C 30
D 35

26 The club sponsor starts packing boxes at this time.

At this time she must leave to pick up her son.

How much time does the sponsor have to work?

F 1 hour and 15 minutes
G 1 hour
H 1 hour and 30 minutes
J 2 hours

27 The club picked and shucked the corn in one day. They began working at 8:00 in the morning. They finished at 1:30 that afternoon. How many hours did they work?

A $9\frac{1}{2}$ **C** $4\frac{1}{2}$

B $3\frac{1}{2}$ **D** $5\frac{1}{2}$

28 The club thinks they can grow twice as much popcorn next year. Which of these is the best estimate of how many jars that would be?

F 180 **H** 200
G 140 **J** 100

This diagram shows the field where the club grew the corn. They only used the shaded section. Study the diagram. Then do Numbers 29 and 30.

corn acres		

29 What fraction of the field did the girls club use?

A $\frac{1}{3}$ **C** $\frac{1}{5}$

B $\frac{1}{4}$ **D** $\frac{1}{6}$

30 Which expression shows how many acres of land they will use if they plant twice as much corn next year?

F A − 2
G A + 2
H A + A
J A ÷ 2

This graph shows which states lie along the gulf coast. The bars show how many miles of coast lie within each state. Study the graph. Then do Numbers 31 through 34.

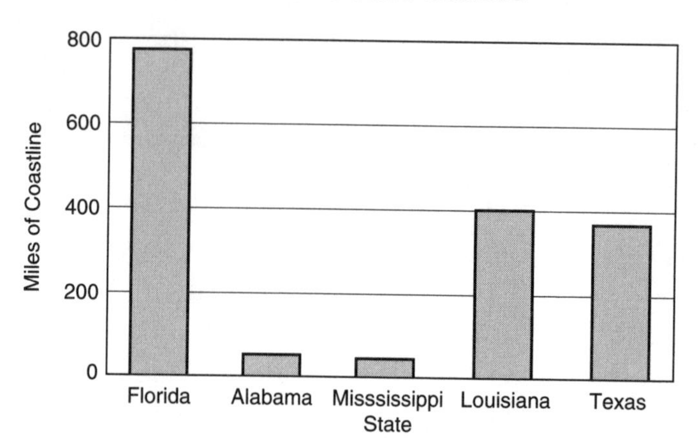

31 About how many miles of coastline does Florida have?

A 650
B 775
C 825
D 625

32 Together, Louisiana and Texas have about how many miles of coastline?

F 600
G 500
H 800
J 1,000

33 Compared to Louisiana, Mississippi has

A much more coastline.
B much less coastline.
C about the same amount of coastline.
D slightly more coastline.

34 Every three days, the coast guard collects water samples at Glendow Beach. They will take a sample April 17. Which of these shows the remaining days in April when the Coast Guard will collect samples?

F 20, 23, 26, 29
G 19, 21, 23, 25, 27, 29
H 21, 25, 29
J 20, 24, 27, 30

This diagram shows where Carl will plant the daisies and the tulips in his flowerbed. He will plant one flower in each square. That is 21 flowers altogether. Study the diagram. Then do Numbers 35 through 39.

35 So far, Carl has planted 7 flowers. How much of the flower bed is planted?

A $\frac{1}{7}$ **C** $\frac{7}{21}$

B $\frac{1}{2}$ **D** $\frac{21}{7}$

36 Daisies cost $2.19 per plant, and tulip bulbs cost $1.95 each. About how much did Carl spend to buy all 21 plants?

F $40.00
G $20.00
H $30.00
J $10.00

37 In which of these number sentences is S the number of squares Carl has created in this diagram?

A $3 + 7 = S$
B $7 \times 3 = S$
C $21 - 7 = S$
D $21 \div 3 = S$

38 This diagram shows a border Carl wants to put along the front of the flower bed.

He knows that there are 7 squares along the front of the bed. What else must he know to find the total length of the border?

F the length of each side of the squares
G how tall the border will be
H how deep the soil is
J how many squares there are along the side of the bed

39 It takes Carl half an hour to plant one row of flowers. How long will it take him to plant all three rows?

A 45 minutes
B 1 hour
C 3 hours
D $1\frac{1}{2}$ hours

On the second Tuesday of December, the business club always holds a breakfast for charity. This year, Boris is in charge of getting the bagels. He must buy 200 bagels. Use this information to do Numbers 40 through 44.

40 What date on this calendar page is the second Tuesday of December?

December

Sun.	Mon.	Tues.	Wed.	Thurs.	Fri.	Sat.
	1	2	3	4	5	6
7	8	9	10	11	12	13
14	15	16	17	18	19	20
21	22	23	24	25	26	27
28	29	30	31			

F December 2
G December 9
H December 16
J December 8

41 There are 6 people on the planning committee. Two of them missed the last meeting. What fraction of the planning committee was absent?

A $\frac{1}{4}$ **C** $\frac{6}{2}$

B $\frac{1}{2}$ **D** $\frac{2}{6}$

42 The bagels cost $0.49 each. About how much will it cost to buy 200 bagels?

F $200.00
G $100.00
H $150.00
J $250.00

43 Boris can buy small packets of cream cheese for $0.25 each. How many packets can he buy for $10.00?

A 40
B 25
C 250
D 200

44 The business club sells 213 tickets to the breakfast. They end up feeding 58 people more than that. How many people did they feed?

F 261
G 271
H 145
J 258

45 A piece was cut out of the folded paper. Then the piece was unfolded. Which of these shows the piece that was cut out?

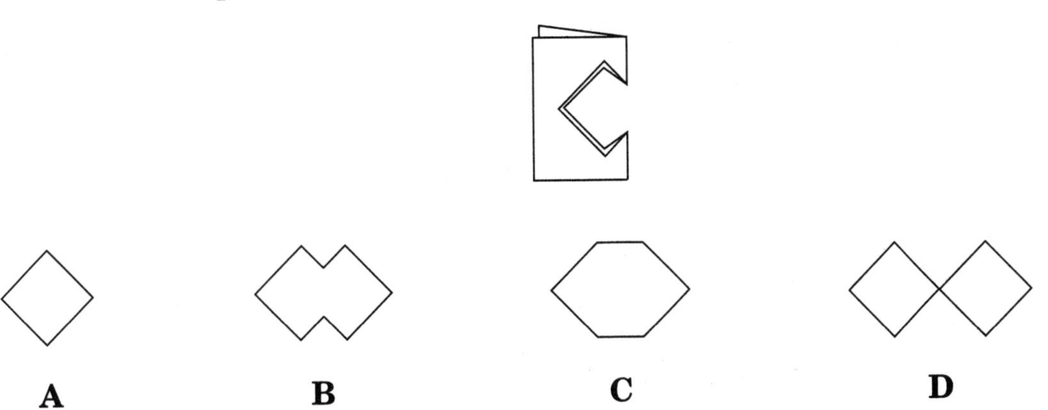

46 Which dotted line shows where to fold the figure so that the parts match?

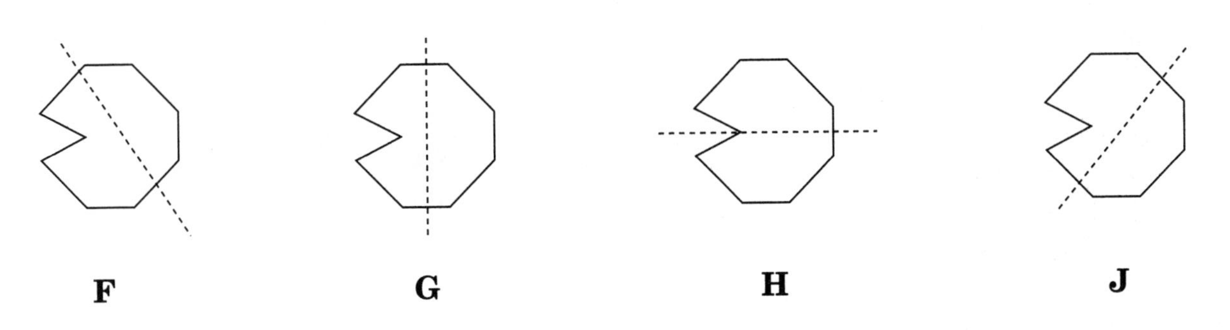

47 Which of these shapes do you make if you trace around the bottom of the object below?

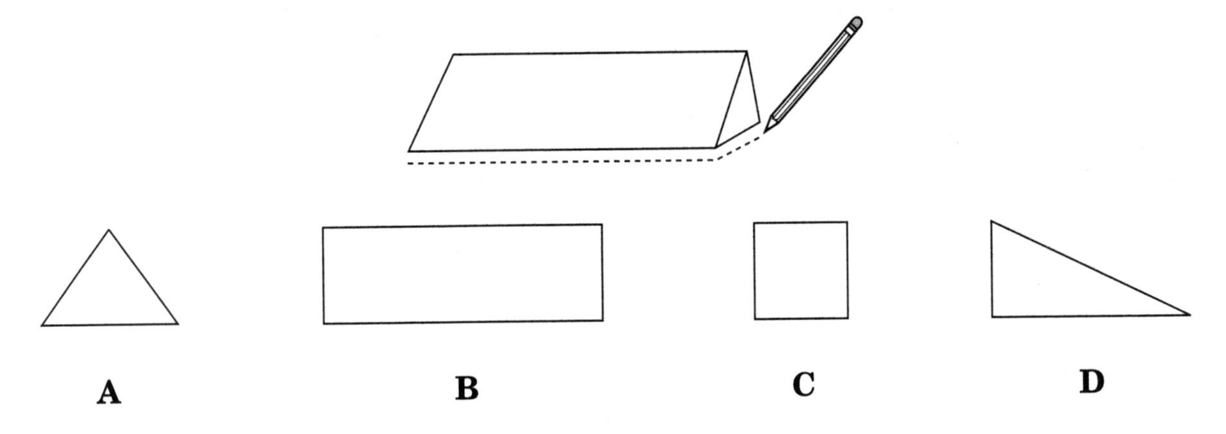

48 What figure is missing from the pattern in the box?

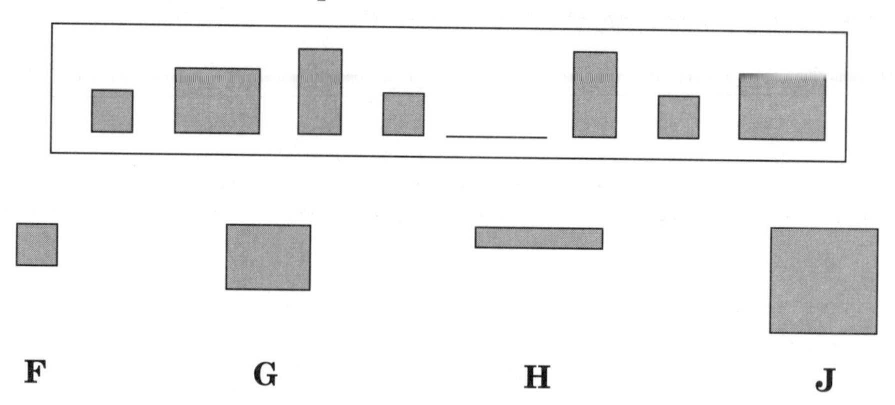

F G H J

49 Which shape has the same size and shape as the figure in the box?

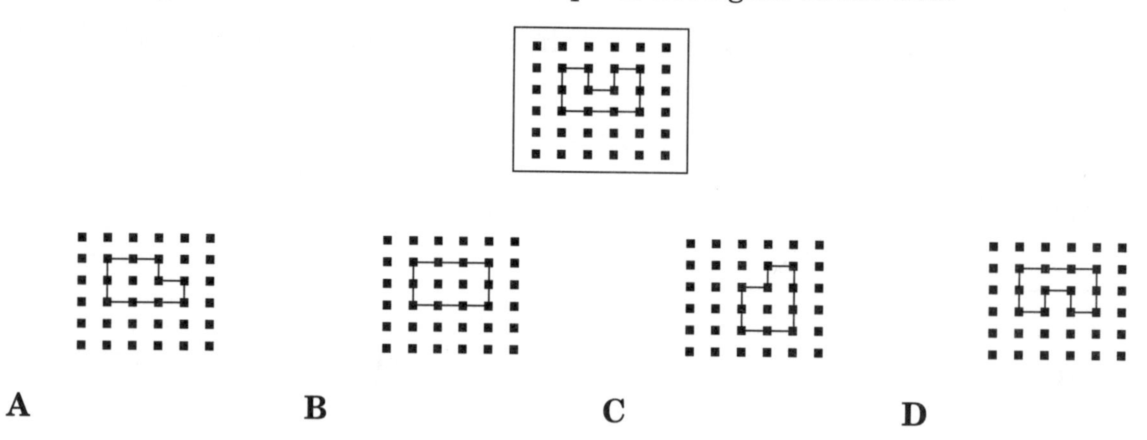

A B C D

50 How many centimeters long is the pencil?

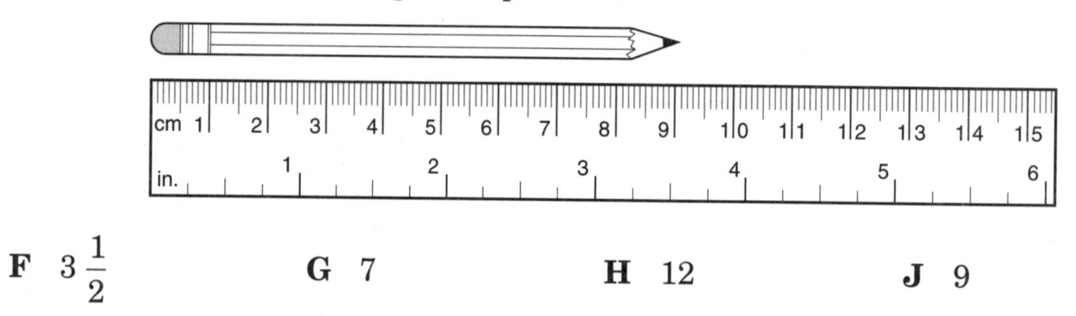

F $3\frac{1}{2}$ **G** 7 **H** 12 **J** 9

Skills Inventory Post-Test Evaluation Chart

Use the key to check your answers on the Post-Test. The Evaluation Chart shows where you can turn in the book to find help with the problems you missed.

Keys

Part A

1	A	14	G
2	K	15	D
3	A	16	F
4	F	17	A
5	E	18	G
6	J	19	D
7	B	20	H
8	J	21	B
9	E	22	J
10	F	23	C
11	B	24	J
12	H	25	E
13	D		

Part B

1	A	26	F
2	H	27	D
3	D	28	F
4	H	29	D
5	C	30	H
6	G	31	B
7	C	32	H
8	J	33	B
9	A	34	F
10	F	35	C
11	C	36	F
12	J	37	B
13	B	38	F
14	G	39	D
15	D	40	G
16	J	41	D
17	B	42	G
18	H	43	A
19	C	44	G
20	G	45	B
21	A	46	H
22	J	47	B
23	C	48	G
24	J	49	D
25	A	50	J

Evaluation Charts

Part A

Problem Numbers	Skill Areas	Practice Pages
1, 2, 3, 10, 11, 12, 13	Addition	17–30
4, 5, 6, 16, 17, 18, 19	Subtraction	31–45
7, 8, 9, 20, 21, 22	Multiplication	46–57
14, 15, 23, 24, 25	Division	58–72

Part B

Problem Numbers	Skill Areas	Practice Pages
1, 3, 5, 7, 10, 23, 29, 35, 42	Numeration/ Number Theory	1–16, 17, 31, 46, 58
8, 11, 12, 13, 18, 31, 33, 40	Data Interpretation	73–85
6, 14, 15, 30, 34, 37	Pre-Algebra	86–99
2, 22, 26, 27, 50	Measurement	100–118
4, 45, 46, 47, 48, 49	Geometry	119–126
9, 16, 17, 20, 24, 38, 39, 43, 44	Computation in Context	8, 26–28, 41–43, 53–55, 68–69
19, 21, 25, 28, 32, 36, 41	Estimation	11–14, 25, 39–40, 52, 67

Answer Key

Pages 1–2, Place Value

1. 9, 5, 3
2. tens, ones
3. 3, 0
4. 4 tens and 5 ones
5. 4 hundreds, 1 ten, and 5 ones
6. 2 hundreds, 0 tens, and 1 one
7. 1 thousand, 2 hundreds, 4 tens, and 6 ones
8. tens
9. thousands
10. ones
11. hundreds
12. 5
13. 7
14. 0
15. three hundred *or* 300
16. ten *or* 10
17. sixty *or* 60
18. four hundred *or* 400

Page 3, Naming Large Numbers

1. 300 + 10 + 9
2. 1,000 + 600 + 40 + 9
3. 2,000 + 60 + 1
4. five hundred forty-six
5. six hundred one
6. four thousand, one hundred thirty
7. one thousand, forty-one
8. three thousand, nine

Page 4, Writing Large Numbers

1. 700 + 6
2. 100 + 40
3. 2,000 + 10 + 3
4. 302
5. 160
6. 1,050
7. 3,006
8. 4,100
9. 10,100
10. 5,306
11. 6,450

Page 5, Comparing Numbers

1. 54
2. 20
3. 99
4. 29
5. 432
6. 50
7. 13
8. 53
9. 29
10. 13
11. 12, 83, 105, 156
12. 7, 10, 59, 100
13. 23, 51, 65, 75, 87
14. 158
15. 359

Page 6, Special Types of Numbers

1. 6, 18, 32, 4, 12, 48, 8
2. 5, 11, 7, 21, 7, 1, 45, 13
3. cat
4. 14th
5. scissors
6. 13th

Page 7, Fractions

1. $\frac{1}{2}$
2. $\frac{1}{4}$
3. $\frac{1}{3}$
4. $\frac{1}{5}$
5. $\frac{2}{3}$
6. $\frac{3}{4}$
7. $\frac{2}{6}$ *or* $\frac{1}{3}$
8. $\frac{2}{4}$ *or* $\frac{1}{2}$
9. The rectangle should be divided into three sections. Two sections should be shaded.
10. The rectangle should be divided into four sections. One section should be shaded.
11. The rectangle should be divided into five sections. Two should be shaded.
12. The rectangle should be divided into four sections. Three should be shaded.

Page 8, Mixed Numbers

1. $2\frac{1}{2}$
2. $2\frac{1}{4}$
3. $1\frac{5}{8}$
4. $1\frac{1}{2}$
5. $\frac{2}{7}$
6. $\frac{5}{10}$ *or* $\frac{1}{2}$
7. $\frac{5}{12}$
8. $\frac{2}{9}$
9. $\frac{30}{125}$ *or* $\frac{6}{25}$
10. $\frac{2}{10}$ *or* $\frac{1}{5}$
11. $\frac{200}{500}$ *or* $\frac{2}{5}$

Page 9, Using Coins

1. 5
2. 10
3. 5
4. 2
5. 2 and 1 *or* 1 and 3 *or* 0 and 5
6. $\frac{8}{100}$ *or* $\frac{2}{25}$

7. $\frac{15}{100}$ *or* $\frac{3}{20}$

8. $\frac{50}{100}$ *or* $\frac{1}{2}$

9. $\frac{75}{100}$ *or* $\frac{3}{4}$

10. $\frac{20}{100}$ *or* $\frac{1}{5}$

11. $\frac{5}{100}$ *or* $\frac{1}{20}$

12. $\frac{50}{100}$ *or* $\frac{1}{2}$

13. $\frac{10}{100}$ *or* $\frac{1}{10}$

Page 10, Writing Money in Decimal Form

1. three dollars and twenty-three cents
2. four dollars and three cents
3. one dollar and thirteen cents
4. seventy cents *or* zero dollars and seventy cents
5. $4.10
6. $3.05
7. $1.50
8. $2.06
9. $0.32
10. $0.75
11. $0.03
12. $0.25

Page 11, Estimation

1. B
2. F
3. A
4. G
5. A, C, D and E should be circled.

Page 12, Common Sense Estimation

1. B
2. H
3. D
4. H
5. A
6. G

Pages 13–14, Rounding

1. 20
2. 20¢
3. 50
4. 1
5. 110
6. $1.00
7. 1
8. 10
9. 10
10. 20
11. 140
12. 60¢
13. 1 dollar *or* $1.00 *or* 100¢
14. 120
15. 150
16. 280
17. 500
18. 700
19. 1,000
20. 2,000
21. 350
22. 600
23. 1,300

Pages 15–16, Number System Skills Practice

1. B
2. J
3. D
4. H
5. A
6. F
7. B
8. J
9. C
10. J
11. D
12. F
13. D
14. J

Pages 17–18, Basic Concepts

1. A, B, C, F, G
2. T
3. F
4. F
5. F
6. T
7. T

Page 19, Basic Addition Facts

1. 5, 6, 1, 14
2. 16, 10, 4, 15
3. 2, 11, 9, 4
4. 7, 11, 0, 5
5. 7, 15, 8, 17, 18, 4
6. 7, 12, 8, 6, 3, 3
7. 6, 8, 8, 12, 12, 10
8. 10, 11, 2, 9, 9, 5
9. 13, 10, 15, 7, 8
10. 13, 12, 16, 6, 13

Page 20, Adding Three or More Numbers

1. 6, 13, 7, 15, 17, 4
2. 9, 9, 9, 10, 8, 10
3. 12, 7, 9
4. 18, 11, 16
5. 10, 15, 16

Page 21, Adding in Column Form

1. 54, 581, 73, 239, 88
2. 95, 48, 679, 98, 89
3. 974, 99, 74, 609, 29
4. $91, $5.08, 53 inches, $4.60, 68 miles

Page 22, Adding Small and Large Numbers

1. 74, 546, 198, 99, 28 feet
2. 49, 628, 87, 329, $79
3. 17
4. 29
5. $19
6. 549
7. 361
8. 936

Pages 23–24, When a Column Sum Is More than 9

1. 52, 108, 92, 22, 81
2. 31, 60, 118, 54, $32
3. 151, 611, 147, 40 inches, 80
4. 212, 60, 91, 28, 514
5. 104, 115, $90.00, 26, 109
6. 51, 100, 90, 117, 27 cups
7. 23
8. 70
9. 41
10. 110
11. 162
12. 45
13. 122
14. 35
15. 701
16. 614
17. 100
18. 131
19. 429
20. 105

Page 25, Using Estimation To Check Addition

1. 70, 50, 50 yards, $90
2. 600; 1,100; 700; 500
3. D
4. H
5. D
6. H

Pages 26–28, Solving Word Problems

1. A, C, D, G, and I should be circled. All the other problems call for subtraction.
2. 16 + 14 + 2 = 32
3. $150 + $70 = $220
4. 8 cents + 15 cents = 23 cents
5. 3 + 4 = 7
6. 54 + 12 = 66
7. how much the groceries cost
8. what time it is
9. how much the parts cost
10. how many albums Anthony sold
11. 18
12. 20

Pages 29–30, Addition Skills Practice

1. B
2. H
3. D
4. F
5. A
6. J
7. C
8. F
9. D
10. K
11. C
12. F
13. B
14. H
15. D
16. H
17. B
18. J
19. E

Pages 31–32, Basic Concepts

1. B, C, D, and F
2. 61
3. 56
4. 107 − 32 = 75 *or* 107 − 75 = 32
5. 133 + 56 = 189 *or* 56 + 133 = 189
6. 5 − 2
7. 5 − 3
8. F
9. F
10. T
11. T
12. T
13. F
14. T

Page 33, Basic Subtraction Facts

1. 2, 4, 1, 5
2. 0, 2, 8, 3
3. 9, 1, 3, 2
4. 3, 3, 1, 6
5. 1, 1, 0, 1, 6, 4
6. 7, 2, 2, 2, 1, 3
7. 4, 4, 6, 4, 1, 7
8. 6, 5, 8, 7, 1, 5
9. 5, 0, 3, 5, 8
10. 3, 6, 2, 4, 4

Pages 34–35, Subtracting in Column Form

1. 53, 122, 321, 384, 63
2. 19, 30, 306, 35, 11
3. 122
4. 27
5. 42
6. 141
7. $49
8. $3.60
9. $1.02
10. 46 feet
11. 7
12. 50
13. $2.00
14. 1 ft
15. 63
16. 114
17. 73 in.
18. 222
19. $1.26
20. 550 meters
21. 42
22. 110
23. 402

Pages 36–37, How To Borrow

1. 28
2. 76
3. 57
4. 2
5. 29 votes
6. 5 apples
7. 66
8. 25
9. 38
10. 362
11. 485 pounds
12. 143
13. 745 cups
14. 350

15. 594
16. 442
17. 339
18. 368
19. 7,157

Page 38, Borrowing from Zero

1. 494
2. 788
3. 684
4. $58
5. 931
6. 473 ounces
7. 137
8. 2,329
9. 91
10. 365
11. 488
12. 796

Page 39, Using Estimation To Check Subtraction

1. 80 – 50 = 30
2. 80 – 20 = 60
3. 700 – 700 = 0
4. $60 – $40 = $20
5. 100 – 90 = 10
6. 500 feet – 90 feet = 410 feet
7. $4.00 – $2.00 = $2.00
8. $2.00 – $1.00 = $1.00
9. 900 – 300 = 600
10. 400 – 40 = 360
11. 700 – 50 = 650
12. 1,000 – 500 = 500

Page 40, Checking Subtraction with First Digits

1. 50; less than 50
2. 80 lb; more than 80 lb
3. 500; more than 500
4. 100; less than 100
5. 200; less than 200
6. C
7. F
8. B
9. H

Pages 41–43, Solving Word Problems

1. B
2. J
3. A
4. H
5. $290 + $60 = $350
6. 183 + 17 = 200
7. $830 – $550 = $280
8. $4.10 – $0.20 = $3.90
9. how many miles she has driven so far *or* her average speed
10. how many tickets he bought *or* how many people are in his family
11. how many have high school or other degrees
12. how much his younger brother makes
13. 25
14. $315

Pages 44–45, Subtraction Skills Practice

1. D
2. F
3. B
4. H
5. B
6. G
7. E
8. J
9. A
10. H
11. E
12. H
13. A
14. G
15. B
16. F
17. B
18. F
19. E
20. F

Pages 46–47, Basic Concepts

1. 262
2. 0
3. 15 × 4 *or* 4 × 15
4. 61 + 61 + 61
5. 4 × 3 *or* 2 × 6
6. T
7. F
8. T
9. F
10. F
11. F
12. T
13. T

Page 48, Basic Multiplication Facts

1. 12, 16, 8
2. 6, 32, 36
3. 30, 15, 42
4. 10, 28, 72
5. 12, 4, 21, 27
6. 49, 24, 9, 20
7. 45, 35, 14, 24
8. 16, 18, 40, 64
9. 36, 56, 25, 48
10. 63, 18, 81, 54

Page 49, Multiplying by a One-Digit Number

1. 69, 282, 242, 168, 217
2. 309, 159, 70, 128, 328
3. 268
4. 108
5. 320
6. 630

Pages 50–51, When a Column Product Is More Than 9

1. 60
2. 81
3. 138
4. 108
5. 112
6. 50
7. 225
8. 84
9. 105
10. 207 pounds
11. 81 dollars
12. 90 inches
13. 111 feet
14. 945

15. $621
16. 1,200
17. 1,683
18. 1,410
19. 2,800
20. 625
21. 411
22. $441
23. 468
24. 510
25. 948 feet
26. 1,300
27. 2,520
28. 378

Page 52, Using Estimation When You Multiply

1. $80 \times 2 = 160$
2. $90 \times 9 = 810$
3. $30 \times 9 = 270$
4. $\$100 \times 3 = \$300$
5. $200 \times 6 = 1{,}200$
6. $600 \times 8 = 4{,}800$
7. $\$400 \times 5 = \$2{,}000$
8. $7{,}000 \times 2 = 14{,}000$
9. $900 \times 3 = 2{,}700$
10. $300 \times 9 = 2{,}700$
11. $2{,}000 \times 6 = 12{,}000$
12. $1{,}000 \times 7 = 7{,}000$

Pages 53–54, Solving Word Problems

1. A
2. G
3. C
4. H
5. C
6. G
7. $\$515 + \$30 = \$545$
8. $146 - 135 = 11$
9. $6 \times \$25 = \$150$
10. $\$30 - \$25 = \$5$
11. $\$1.59 \times 9 = \$14.31$
12. $\$1.59 - \$1.29 = \$0.30$
13. about how many apples are in each box
14. about how far he hikes in one day
15. how many classes there are
16. 64 dollars
17. 555 feet
18. 42 miles per hour
19. 48 racks

Pages 56–57, Multiplication Skills Practice

1. B
2. F
3. C
4. G
5. B
6. H
7. D
8. F
9. C
10. G
11. A
12. F
13. D
14. H
15. A
16. J
17. C
18. G
19. C

Pages 58–59, Basic Concepts

1. A and D should be circled.
2. 36
3. 0
4. 42
5. 75
6. 1
7. $4\overline{)16}$
8. $93 \div 3 = 31$ *or* $93 \div 31 = 3$
9. $56 \times 4 = 224$ *or* $4 \times 56 = 224$
10. F
11. F
12. F
13. F
14. F

Page 60, Basic Division Facts

1. 2, 4, 2
2. 2, 8, 6
3. 6, 5, 3
4. 5, 7, 0
5. 4, 2, 7, 9
6. 7, 8, 7, 4
7. 9, 7, 2, 4
8. 8, 9, 8, 8
9. 9, 7, 5, 6
10. 7, 6, 9, 6

Page 61, Using the Division Symbol $\overline{)}$

1. $2\overline{)64}$
2. $8\overline{)88}$
3. $3\overline{)93}$
4. $5\overline{)55}$
5. $3\overline{)63}$
6. $4\overline{)28 \text{ inches}}$
7. $4\overline{)48}$
8. $2\overline{)28}$
9. $3\overline{)39}$
10. $4\overline{)84 \text{ dollars}}$
11. $2\overline{)\$46}$

Page 62, Dividing by a One-Digit Number

1. 31
2. 11
3. 42
4. 20
5. 121
6. 112
7. 14
8. $20
9. 23
10. 32
11. 21
12. $50

Page 63, When a Dividend Digit Is Too Small

1. 53
2. 81
3. 21
4. 50

5. 51
6. 62
7. 74
8. 20
9. 105
10. 108
11. 56
12. 205
13. 105
14. 104
15. 305
16. 102

Page 64, Dividing with a Remainder

1. 2 r 1
2. 3 r 1
3. 3 r 3
4. 5 r 4
5. 3 r 5
6. 7 r 8
7. 7 r 1
8. 8 r 3
9. 4 r 2
10. 6 r 3
11. 7 coins each, 2 left over

Page 65, Writing the Steps of a Division Problem

1. 18
2. 14
3. 15
4. 28
5. 18
6. 12
7. 69
8. 121
9. 74
10. 33
11. 29
12. 33
13. 13
14. 15 pieces; 2 inches

Page 66, Mixed Practice

1. 211
2. 9
3. 301
4. 9
5. 36
6. 30
7. 400
8. 102
9. 9
10. 6 r 1
11. 161
12. 305
13. 30
14. 16
15. 66
16. 101 r 2
17. 5 frames
18. 5 feet
19. $41
20. 17

Page 67, Using Estimation with Division

1. 70
2. 100
3. 100
4. 70
5. 600
6. 300
7. 300; exact answer
8. 40; exact answer
9. 120; exact answer
10. 50; exact answer
11. 40; estimate
12. 100; exact answer
13. 80; exact answer
14. 600; exact answer
15. 50; exact answer

Page 68, Solving Word Problems

1. $26,000
2. $100
3. $625
4. $112
5. $25
6. $240
7. $40 per yard
8. $265

Pages 69–70, Solving Two-Step Word Problems

1. B
2. G
3. A
4. H
5. Subtract $7 from $60. Then subtract another $35. Or, add $35 and $7. Then subtract the sum from $60.
6. Subtract 3 yards from 6 yards. Then divide the difference by 3.
7. ~~$36,000~~ $3,000
8. 4 dollars
9. $200
10. $2
11. $74

Pages 71–72, Division Skills Practice

1. B
2. J
3. E
4. J
5. E
6. K
7. A
8. J
9. D
10. K
11. C
12. F
13. D
14. F
15. D
16. J
17. C
18. K
19. B

Pages 73–74, Reading a Table

1. D
2. G
3. A
4. J
5. $1.25
6. ~~$1.45~~ 1.75
7. $3.05
8. a small flavored coffee
9. a medium steamed milk
10. a small latte

Page 75, Using Numbers in a Table

1. December
2. theft
3. December
4. January
5. B

Page 76, Comparing Numbers in a Table

1. Ruby Creek trail
2. the Old Mill trail
3. 10
4. 20
5. 2
6. 6
7. $\frac{5}{15}$ *or* $\frac{1}{3}$
8. $\frac{10}{40}$ *or* $\frac{1}{4}$
9. $\frac{15}{20}$ *or* $\frac{3}{4}$
10. 8
11. $\frac{5}{20}$ *or* $\frac{1}{4}$

Page 77, Using a Calendar

1. Wednesday
2. Sunday
3. 15
4. November 20
5. 4
6. November 11

Page 78, Using a Price List

1. $9.20
2. $9.80
3. $1.50
4. $12.00
5. $14.25
6. 3 pounds
7. $\frac{4}{5}$
8. macaroni and cheese

Page 79, Graphs

1. 4
2. 4
3. January
4. 5
5. 1999

Page 80, Reading a Pictograph

1. 10 million people
2. 21 to 40
3. 70 million
4. 2
5. 5 million people
6. Americans under 21
7. $\frac{40}{250}$ or $\frac{4}{25}$

Page 81, Reading a Bar Graph

1. $80
2. $20
3. $10
4. $140
5. $70
6. February
7. $\frac{60}{80}$ *or* $\frac{3}{4}$

Page 82, Reading a Circle Graph

1. $\frac{2}{10}$ *or* $\frac{1}{5}$
2. F
3. D
4. J
5. D
6. G

Page 83, Finding an Average

1. Jackie 10; Jordon 11; Rene 12; Jamie 11
2. G
3. C

Pages 84–85, Data Interpretation Skills Practice

1. A
2. F
3. B
4. H
5. C
6. J
7. D
8. J
9. C
10. H
11. D
12. J
13. H
14. B

Pages 86–87, Finding Patterns

There should be drawings of each of the following.

1. a square
2. a stack of 5 bars
3. an empty circle
4. a dark star
5. an arrow pointing down
6. a dark triangle
7. a square with the lower right-hand corner shaded
8. C
9. thick
10. B
11. eight
12. A
13. C

Page 88, Finding Number Patterns

1. 8, 10, 12, 14
2. 25, 30, 35
3. 12, 15, 18
4. fours, four
5. 2
6. 5
7. 2
8. counting by sixes, starting with 0 *or* add six to each number

9. counting by fours, starting with one *or* add four to each number
10. subtract ten from each number *or* counting backwards by tens, starting with 90
11. 14
12. 18, 30
13. 27

Page 89, Patterns in Multiplication and Division Facts

1. 2, twos
2. even
3. 5, 0
4. B
5. They increase by one.
6. 81
7. 0; the other number being multiplied
8. 650
9. 5 and 2

Page 90, Patterns in Graphs and Tables

1. \$0.30 *or* 30 cents
2. \$0.20 *or* 20 cents
3. \$1.45
4. \$3.40
5. decrease
6. D

Pages 91–92, Patterns in Number Sentences

1. +
2. –
3. +
4. +
5. –
6. 5
7. 1
8. 3
9. 5
10. 0
11. Add 6.
12. Add 7.
13. Subtract 1.
14. Add 2.
15. Subtract 6.
16. Add 5.
17. Multiply by 2.
18. Subtract 10.

Page 93–94, Writing Letters and Symbols for Words

1. $n + 12$
2. $n - 3$
3. $n + 4$
4. $n + 1$
5. $n - 3$
6. $n \times 2$ *or* $2n$
7. $5 + n$
8. $n \div 6$
9. $3 \div n$
10. $n - 32$
11. $7 \times n$ *or* $7n$
12. $12 + n$
13. $s - 2$
14. $x - 15$
15. $F + S$
16. $c + 2$
17. $4 \times b$ *or* $4b$
18. $12 + c$
19. $96 - b$
20. $3 + g$

Page 95, Writing Equations

1. $3 + 2 = x$
2. $x - 6 = 8$
3. $3 + 4 = x$
4. $x - 7 = 6$
5. $8 - 5 = x$
6. $8 - 2 - 2 = x$

Pages 96–97, Solving Equations

1. 5
2. 2
3. 4
4. 9
5. 3
6. 4
7. 3
8. 13
9. 80
10. 4
11. 11
12. 70
13. $x + 9 = 15, x = 6$
14. $x - 3 = 7, x = 10$
15. $x - 5 = 10, x = 15$
16. $5 - x = 3, x = 2$

Pages 98–99, Pre-Algebra Skills Practice

1. D
2. H
3. B
4. G
5. C
6. G
7. A
8. G
9. B

Page 100, Choosing the Best Tool

1. E
2. C
3. D
4. E
5. H
6. B
7. F
8. C
9. A
10. E

Pages 101–102, Reading a Scale

1. 7
2. 25
3. $3\frac{1}{2}$
4. 75
5. 3
6. 65
7. $\frac{3}{4}$; $3\frac{3}{4}$ or 3.75
8. $1\frac{1}{4}$
9. 10, 80
10. 25, 275
11. 1, 12
12. 150
13. 14

14. 180
15. 37.5

Page 103, Estimating a Measurement on a Scale

1. A
2. G
3. D
4. J
5. B
6. F

Page 104, Measuring Temperature

1. degrees Fahrenheit
2. 32°F
3. colder
4. D
5. 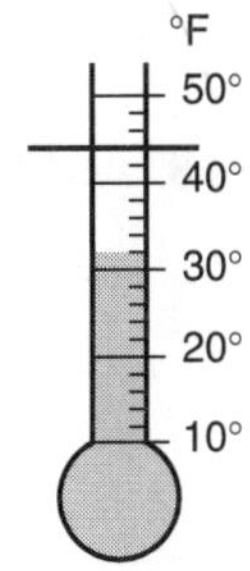

6. one degree Celsius
7. 2°F, 2°C
8. about 37°C; about 100°F
9. B
10. summer
11. 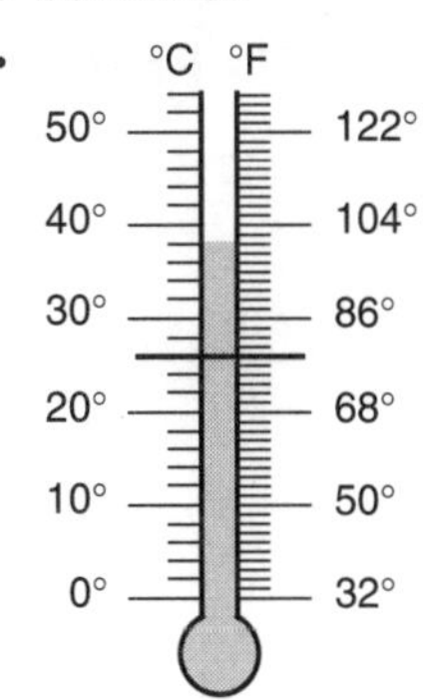

Pages 105–106, Measuring Length

1. B
2. F
3. B
4. F
5. 24 inches
6. 1 yard
7. 6 inches
8. 2 yards
9. $\frac{1}{3}$
10. a little more than 15 cm
11. a little more than 6 in.
12. 10
13. $2\frac{1}{2}$ cm
14. 11 cm
15. $4\frac{1}{2}$ in.
16. $7\frac{1}{2}$ cm
17. 70 mm
18. $1\frac{1}{2}$ in.
19. 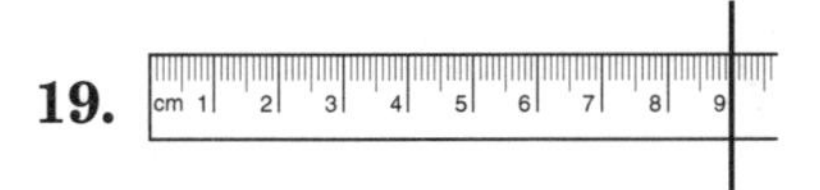

20. D

Page 107, Using a Ruler

1. 4 in.
2. 5 in.
3. 3 in.
4. 6 in.
5. 2 in.
6. 7 in.

Page 108, Reading Fractions on Scales

1. $1\frac{1}{2}$ in.
2. $\frac{1}{2}$ in.
3. 2 in.
4. $2\frac{1}{2}$ in.

5–7. Check book: 6 by 3
computer paper: $8\frac{1}{2}$ by 11
car key: $2\frac{3}{4}$ by $1\frac{1}{4}$
business card: $3\frac{1}{2}$ by 2
cassette tape: 4 by $2\frac{1}{2}$
news magazine: 10 by 7

Page 109, Measuring to the Nearest $\frac{1}{8}$-inch

1. $1\frac{3}{8}$
2. $2\frac{1}{8}$
3. $\frac{7}{8}$
4. $\frac{5}{8}$
5. $1\frac{7}{8}$

For the objects, answers may vary.

Page 110, Measuring to the Nearest Millimeter

1. 35
2. 54
3. 22
4. 16
5. 48

For the objects, answers may vary.

Page 111, Measuring Weight

1. 8 oz
2. 500 g
3. A
4. H
5. $3\frac{1}{2}$ lb
6. 4 kg
7. 3 kg
8. 2 lb
9. 64 oz
10. 2,000 g
11. 1 lb

Page 112, Measuring Liquids

1. 4 cups
2. 8 quarts

3. 2,000 mL
4. 3; $\frac{3}{4}$
5. A
6. B
7. 1
8. $\frac{1}{4}$ cup, under 100 mL
9. $\frac{1}{2}$
10. 250
11. $1\frac{1}{2}$

Page 113, Reading and Writing Time

1. 7:15; seven fifteen; a quarter after seven; fifteen minutes after seven
2. 10:50; ten fifty; ten to eleven; fifty minutes after ten
3. 9:32; nine thirty-two
4. 30 min
5. 15 min
6. 90; 30
7. 2 hr
8. 12 hr

Pages 114–115, Changes in Time

1. 11:00
2. 3:30
3. 6:15
4. 1:30
5. 1:15
6. 1:30
7. 1 hour and 5 minutes
8. 3 hours and 20 minutes
9. 6 hours and 20 minutes
10. 30 minutes (*or* $\frac{1}{2}$ hour)
11. 2 hours and 30 minutes (*or* $2\frac{1}{2}$ hours)
12. 8 hours and 30 minutes (*or* $8\frac{1}{2}$ hours)

Page 116, Reading a Map or Diagram

1. D
2. $4\frac{1}{2}$ miles
3. 2 miles
4. A
5. 4:00
6. 12 minutes

Pages 117–118, Measurement Skills Practice

1. A
2. G
3. D
4. J
5. B
6. F
7. C
8. J
9. D
10. H
11. B
12. G
13. A
14. J

Page 119, Using Logic

1. 1
2. the square should be circled
3. D
4. J
5. C
6. C

Page 120, Figures

1. correct
2. correct
3. incorrect
4. correct
5. incorrect

Pages 121–122, Drawing Figures

1. B
2. G
3. A
4. J
5. D
6. F
7. B
8. G
9. A

Page 123, Figures with the Same Shape

1. A
2. H
3. B
4. H
5. A

Page 124, Figures and Symmetry

1. C
2. J
3. A
4. H
5. B
6. G
7. C

Pages 125–126, Geometry Skills Practice

1. D
2. G
3. D
4. J
5. A
6. G

Glossary of Common Terms

algebra: the study of mathematical expressions in which letters stand for unknown numbers

average: one way of representing the most typical number in a set of numbers. The average (or mean) is the sum of all the values in a set divided by the total number of values. Example: 2, 4, 6, 8
To find the average in this set, add 2, 4, 6, and 8. Then, divide the sum (20) by 4. The average is 5.

bar graph: a graph that uses thick bars to represent numbers

borrowing: taking value from one place in a number so you can subtract from a lower place. Example:

$$\begin{array}{r} {}^{1}\not{2}\,{}^{14}\not{4} \\ -\ 1\ 7 \\ \hline 7 \end{array}$$

capacity: how much an object will hold. Example: Most coffee cups have a capacity of 1 to 2 cups.

carrying: taking value from one column in a problem and adding it to the product or sum of the next column. Examples:

$$\begin{array}{r} {}^{1}1\ 5 \\ +\ 3\ 8 \\ \hline 5\ 3 \end{array} \qquad \begin{array}{r} {}^{2}1\ 8 \\ \times\ 3 \\ \hline 5\ 4 \end{array}$$

circle: all the points that are a particular distance from a center point

circle graph: a graph that shows parts of a whole as slices or wedges in a circle

column: numbers, words, or symbols stacked one over the other

common-sense estimation: estimation based on experience rather than calculation

decimal form (of money): When an amount of money is written in decimal form, a small dot (or decimal) separates dollars from cents. Example: $2.15 is two dollars and fifteen cents.

difference: the result when you subtract

digit: one of the following ten numbers: 0, 1, 2, 3, 4, 5, 6, 7, 8, or 9. The number 145 has three digits, 1, 4, and 5.

dividend: the number in a division problem that is being divided. Example: In the problem $30 \div 6 = 5$, 30 is the dividend.

divisor: the number in a division problem that you are dividing by. Example: In the problem $30 \div 6 = 5$, 6 is the divisor.

equation: a number sentence showing that two values are equal. Example: $x + 7 = 8$

estimate: a number that is close to the exact amount

even number: a number that can be evenly divided by two. Examples: 2, 4, 6, 8, 10

fraction: a way of showing part of something. Fractions contain two numbers. The bottom number tells how many parts there are in the whole. The top number shows how many parts the fraction stands for. Example: If there are 23 people in your whole class and 5 of those people are missing, $\frac{5}{23}$ of the class is absent.

front-end estimation: estimating the answer to a problem by using only the first digit in each number

geometry: the study of points, lines, angles, shapes, and solids

inverse operations: operations that are the opposite of each other. Example: Addition is the inverse of subtraction. In other words, the act of adding 12 can be reversed or undone by subtracting 12.

key: a table that explains the meaning of symbols, words, or numbers used in a map or a graph

line graph: a graph that uses dots to show numbers. Lines connect the dots to show how values rise and fall.

mean: *See* average.

metric units: the units of length, capacity, and weight used in Canada and many other parts of the world. Examples: centimeter, liter, gram

mixed number: a combination of a whole number and a fraction. A mixed number represents a value between two whole numbers. Example: $2\frac{1}{2}$

multiples of a number: numbers that can be evenly divided by that number. Example: 9, 15, 21, and 30 are all multiples of 3.

odd number: a whole number that cannot be evenly divided by two. Examples: 1, 3, 5, 7

ordinal number: a number name that shows the place of something in a list. Examples: first, second, third

pictograph: a graph that uses pictures or symbols to represent numbers

place: the location of a digit in a number. Example: 173 has three places. The digit 1 is in the hundreds place, 7 is in the tens place, and 3 is in the ones place.

product: the result when you multiply

quadrilateral: any figure with four sides

quotient: the result when you divide

remainder: the "left over" portion in division problems that can't be solved with a whole number. Example: 16 divided by 5 is 3 with a remainder of 1. In other words, 1 is the difference between 16 and (5×3).

rounding: increasing or decreasing a number so that it ends in one or more zeros. Example: 189, rounded to the nearest 100, is 200.

scale: a number line on a measurement tool

signal word: a term in a word problem that suggests how the problem should be solved

square: a figure with four equal sides and four equal angles

standard units: the units of length, capacity, and weight used in the United States. Examples: inch, pound, cup

sum: the result when you add

table: information organized in columns and rows

tick marks: evenly spaced lines on a scale

total: the result when you add

triangle: a 3-sided figure

unknown: a letter (or variable) in an algebra expression. Unknowns take the place of numbers that haven't been identified. Example: In the expression $2a + 3 = 11$, a is the unknown.

variable: *See* unknown.

Index